Introduction to Focused Ion Beam Nanometrology

Introduction to Focused Ion Beam Nanometrology

David C Cox

National Physical Laboratory, Teddington, UK
and
Advanced Technology Institute, University of Surrey, Guildford, UK

Morgan & Claypool Publishers

ISBN   978-1-6817-4084-3 (ebook)
ISBN   978-1-6817-4020-1 (print)
ISBN   978-1-6817-4212-0 (mobi)

DOI   10.1088/978-1-6817-4084-3

Version: 20151001

IOP Concise Physics
ISSN 2053-2571 (online)
ISSN 2054-7307 (print)

A Morgan & Claypool publication as part of IOP Concise Physics
Published by Morgan & Claypool Publishers, 40 Oak Drive, San Rafael, CA, 94903, USA

IOP Publishing, Temple Circus, Temple Way, Bristol BS1 6HG, UK

To Lúcia, and my parents.

Contents

Preface   ix

Acknowledgements   x

Author biography   xi

1   Metrology   **1-1**

1.1   What is metrology?   1-1
1.2   Metrology in the FIB   1-2
    References   1-6

2   Focused ion beam   **2-1**

2.1   Introduction to the FIB instrument   2-1
2.2   Types of instrument   2-3
    2.2.1   Gas field-ion source   2-5
    2.2.2   Liquid metal ion source   2-6
    2.2.3   Liquid metal alloy ion source   2-8
    2.2.4   Inductively coupled plasma ion source   2-9
2.3   Gas injection systems   2-11
2.4   Patterning options   2-13
2.5   Other equipment and techniques found on FIB instruments   2-15
    References   2-17

3   Ion–solid interactions   **3-1**

3.1   Overview   3-1
3.2   Imaging—secondary electrons and secondary ions   3-2
3.3   Ion milling—ion range, sputter yield and damage   3-5
3.4   Software to approximate ion range, damage and sputter yield   3-11
    References   3-14

4   Focused ion beam—materials science applications   **4-1**

4.1   Overview   4-1
4.2   TEM foils and cross-sectioning   4-1
4.3   Three-dimensional reconstruction   4-3
4.4   Mechanical testing   4-8

4.5   Residual stress measurement and deformation   4-11
4.6   Secondary ion mass spectrometry and atom probe   4-15
      References   4-16

5   Focused ion beam fabrication for metrology   5-1
5.1   Overview   5-1
5.2   Superconducting devices   5-1
5.3   Utilising manipulation systems   5-5
5.4   AFM cantilevers and dimensional artefacts for scanning   5-7
      probe techniques
5.5   Other devices   5-10
      References   5-11

6   Future developments   6-1
6.1   Where we currently are   6-1
6.2   The end of the Ga ion source?   6-2
6.3   Final thoughts   6-3

Preface

When I was invited by Institute of Physics Publishing to write this book a little over a year ago, I went through a process I am sure is common to many writers. After a week of deliberation, I said yes. Some 24 hours later I wondered what was going through my mind when I had said yes. Why on the Earth did I agree to do it? I thought my days of writing large bodies of work started and stopped almost twenty years ago with completion of my PhD thesis. Compiling list of references, sourcing and preparing figures, and sitting for hours on end rewriting the same paragraph over and over and not being happy with it; I must have been mad. However, after one or two false starts I finally got going, and must say, as time progressed have slowly enjoyed the process more and more. It has been made considerably easier by discussions with my colleagues and the manufacturers, who have helped enormously and provided valuable input to this text. Special thanks go to Helen Jones, at NPL, Raphaela Scharfschwerdt at FEI and Diane Stewart at Zeiss.

I would like to thank all my colleagues at the the National Physical Laboratory who over the last ten years have presented me with many scientific challenges, often in areas where I had little or no knowledge. If ever there was an illustration of the power and diversity of the focused ion beam (FIB) it is that in those ten years I have worked with people in almost every division of the laboratory, from quantum metrology, through material science, to dimensional metrology and analytical science.

Indeed, conveying this diversity and the incredible usefulness of these instruments has been my main purpose in writing this book. Firstly, I would like to introduce this wonderful instrument, the FIB, to new and potential users, and to show existing users the additional areas where their instruments may find use. Very few instruments offer such a huge range of diverse applications. Secondly, such is the speed of development of the instruments and associated techniques that although previous (and extremely thorough) texts exist in the subject, they do not cover the latest developments involving new ion sources and methods. Finally, I would like to make the reader aware that some of these methods are not foolproof, that errors exist and are often overlooked. It is only by understanding the limitations of these processes that we truly learn how useful they are. This is not to say that the FIB is a poor technique by any means. Many of the topics covered in this book simply could not be carried out at all without the FIB instrument and many will get even better in terms of both throughput and accuracy as our understanding of them gets better, and creative people develop the instruments and methods further.

David C Cox

Acknowledgements

I would like to thank my many colleagues from the National Physical Laboratory, in particular those who provided material for this book, proof read drafts, or engaged in useful discussions about its content. I would also like to thank those at the University of Surrey who have done similar. Thanks also to my friends at NTT basic research laboratory for providing such a stimulating place to carry out research and allowing me to help you in establishing your new FIB facility. I would also like to thank FEI and Zeiss for materials, some of which I know you worked hard to source, and of course the publishers both at IOP Publishing and Morgan & Claypool who have made this whole process as easy as such a thing can be. Finally thanks to Richard Leach for suggesting I do this project in the first place.

Author biography

David C Cox

David C Cox received his PhD from the department of Metallurgy and Materials Science, University of Cambridge, UK in 2001. He is currently a senior research fellow at the Advanced Technology Institute, University of Surrey, UK, but has been seconded to the National Physical Laboratory (NPL), UK as a senior research scientist since 2005. Having a broad background in industry and academia, covering many aspects of materials science, physics and electronic engineering, he has published close to 100 articles at the time of writing. Largely associated with both the Quantum Metrology and Materials groups at NPL, his most recent research work has concentrated on the area of using the focused ion beam to fabricate devices for quantum metrology. Additionally, he has developed a strong interest in the wider aspects of focused ion beam fabrication and the fundamental understanding of how it can be used to study materials, together with the errors associated with some areas of the technique.

IOP Concise Physics

Introduction to Focused Ion Beam Nanometrology

David C Cox

Chapter 1

Metrology

1.1 What is metrology?

Metrology is the science of measurement and can have several meanings, depending on whom or what it is being applied to. Broadly, these can be classified in three areas. Firstly, it is used to create legal definitions of quantities such as weights and measures, applied to our every day purchases of items and the environment around us. Secondly, all industrial activity is underpinned by metrology. Components manufactured in different parts of our increasingly globalised world must be compatible with components made and assembled elsewhere. These compatibilities might comprise more than simple dimensional agreement, but may also be dependent on other measures such as voltage or chemical composition. Thirdly, fundamental metrology is concerned with developing new methods of measurement, establishing agreed standards, definitions and units of measurement and providing traceable measurements from which standards can be created and applied. It is fundamental metrology that underpins all of the other metrological activities. The field of fundamental metrology is extremely broad, often complex, and sometimes quite abstract. A 2004 definition stressing the huge range to which metrology can be applied was offered by the International Bureau of Weights and Measures (BIPM) 'The science of measurement, embracing both experimental and theoretical determinations at any level of uncertainty in any field of science and technology' [1].

All scientific disciplines have their own well-developed, *and mostly agreed*, use of language and terminology and in this respect metrology is no different. An example of this in the metrological context is the use of the word 'uncertainty' in the preceding paragraph. One thing common to all metrologists is that when we discuss our measurements we tend not to emphasise the absolute values we measure, but we stress the uncertainty (or error) of our measurement as we strive to improve the

doi:10.1088/978-1-6817-4084-3ch1 1-1

science of the measurement itself. We can never be certain of our measurements except in very special cases, such as being certain that no electrical current will flow in an open circuit. As soon as we close the circuit and begin to take measurements we must take into account the precision, accuracy, reliability and (un)certainty associated with taking the measurement [2]. Beginning with precision, we could think of this in simple terms, such as how many decimal places do we have on our meter? However, this approach is incorrect and in fact our precision is also closely linked to how reproducible our measurements are. If we find we take many measurements of the same thing and they agree to three decimal places very well on a system capable in principle of measuring more, we can be confident of only the three decimal place measurement and this determines our precision, with the variation below this contributing to our uncertainty. Turning to accuracy, even if our measurement system agrees very well on identical measurements it does not necessarily translate that it is accurate. To determine accuracy we need to compare our instrument readings to a standard, traceable to one of the national measurement institutes. Reliability is closely linked to precision, but it is also a measure of the accuracy over time. Finally, the uncertainty of our measurement is a product of all of these things and as we make measurements we see variation from sample to sample, day to day and instrument to instrument. This dispersion in the measurements is our measurement uncertainty, with the relative uncertainty being given by the measurement uncertainty divided by the measured value. For readers for whom these concepts are new, a downloadable collection of measurement guides including an introduction to measurement uncertainty can be found at [3].

Throughout this book many examples will be given, or referenced, that quote values and units of measurement. However, the main aim of this text is not to emphasise or assign great importance to these values, but to make the reader aware of the measurement possibilities of using the focused ion beam (FIB) and the likely sources of error and uncertainty and the limits of this very versatile range of instruments.

1.2 Metrology in the FIB

Based on these metrological definitions and key terms, what exactly do we mean by nanometrology? If one thinks of dimensional metrology, and applies the standard SI units and prefixes, a simple definition is the measurement of features from 1×10^{-9} m up to 1×10^{-7} m, this range being in agreement with commonly held definitions of what constitutes a nanoparticle, for example. However, we can use FIB to aid us in the measurement of many other properties, and as we shall see in chapter 4, when using FIB for the measurement of residual stress it is possible to determine stress levels of 1×10^{9} Pascal, 18 orders of magnitude from nano! Limiting ourselves only to dimensional metrology is also complicated, for example, when we produce a transmission electron microscope (TEM) sample with FIB and analyse it in a state-of-the-art TEM we can easily resolve columns of atoms and their spacing, some two orders of magnitude smaller than a nm. This could be argued to be metrology carried out in the TEM, of course, but the sample was still prepared by FIB and it is often the best sample preparation method to enable the measurement to be undertaken. For these reasons, when we refer to nanometrology in the FIB we do mean dimensional metrology on the scale of nm, but we can extend our definition to include using the ion beam to either image, expose, or create features or structures that are on the scale of nm. At the upper limits, some of these things will extend beyond 1×10^{-7} m, and may in one or more dimensions be tens of μm, but the FIB instrument is more than capable of working at the nm length scale and is commonly found working in this regime.

As we have alluded to, practising metrology with focused ion beam instruments is complicated by their high degree of versatility. We can use the instrument to make a direct measurement, we can use it to prepare a sample for measurement in another instrument, or we could even use it to make or modify a completely new instrument or component that can in turn be used to measure something completely unrelated to FIB (figure 1.1). Even in the first case the range of measurements we can make is large due to the varieties of this type of instrument. For example, few FIB systems these days are supplied without a scanning electron microscope (SEM) column, further increasing our measurement capability. However, a detailed discussion on scanning electron microscopy is beyond the scope of this book, where the reader need only be aware that very large numbers of FIB systems also incorporate SEM columns sharing the same vacuum, detectors and sample handing system. We will only discuss electron beam microscopy where necessary to assist in the discussion of the measurement taking place. For an excellent introduction to the SEM and SEM-based techniques, the reader is directed to [4]. Additional third-party-vendor equipment can also take these instruments beyond simple imaging by the addition of x-ray detectors and electron backscatter diffraction (EBSD) cameras. However, while both x-ray emission and backscatter diffraction signals are normally generated by electron beam, the FIB is used to produce the polished faces and slices for both techniques and the preparation of these will be covered in later sections. Finally, the long established and historic use of Ga as the source of ions is being added to with noble gas sources using light elements, He and Ne, largely for imaging, and additionally heavier elements from plasma-based sources such as Xe. These plasma sources offer significantly larger volume ion milling, owing to their high emission currents and the high sputter yield of the heavy ions, bringing FIB into new areas previously deemed beyond the scope of Ga source instruments. Similarly, although currently less common, metallic alloy source based instruments are available, often combining both light and heavy element options in the same instrument.

We can consider three distinct length ranges in FIB microscopy, the size of the sample to be studied, the size of the field of view of our interest in the sample and the resolution at which we can obtain information from the region of interest. Samples studied in the FIB can range from only nm in the case of dispersed nanoparticles, up to 150 mm diameter at the larger end of most laboratory systems. In figure 1.1 we can see that the typical length scale over which a focused ion beam might be used covers some five orders of magnitude, from nm through to hundreds of micrometers, depending on the type of instrument and application. Although the maximum field of view can approach one mm, the instrument is likely to be used to resolve or expose detail with a resolution of a few nm on areas up to around 100 μm at the upper end. As an example, the single most common application for FIB is as a tool to reveal the internal structure of samples. In cases such as this we may be removing tens of μm of material, but we are only interested in imaging a few nm of a thin film to check for uniformity. For example, figure 1.2 shows a simple cross-section cut into a gallium nitride (GaN) film deposited on silicon (Si) wafer.

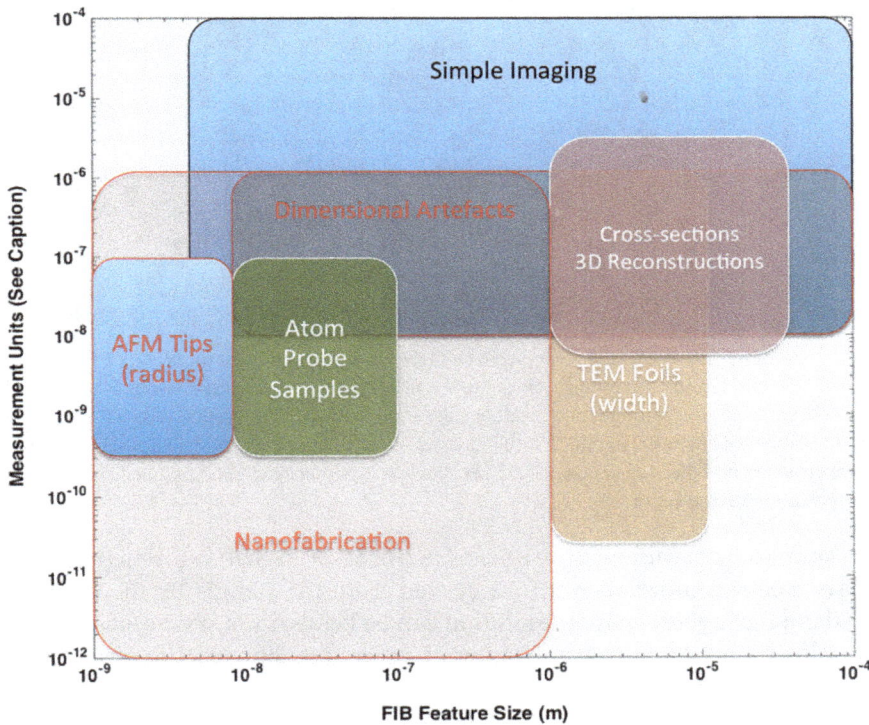

Figure 1.1. Selected examples of the length scales applicable to focused ion beam based metrology. The horizontal axis indicates a typical scale of the feature created or imaged by FIB. The vertical axis indicates the likely range of measurement activity. In most cases this is a simple length scale in metres, but the nanofabrication activities in particular may create devices where other properties are measured, such as magnetic flux or temperature. Text and boxes in black indicate an activity carried out directly in the FIB instrument, white denotes an activity where samples are prepared in a FIB and then measured in another instrument (or with an SEM column on the instrument) and red denotes an activity where devices or artefacts are created in FIB but used for other metrology.

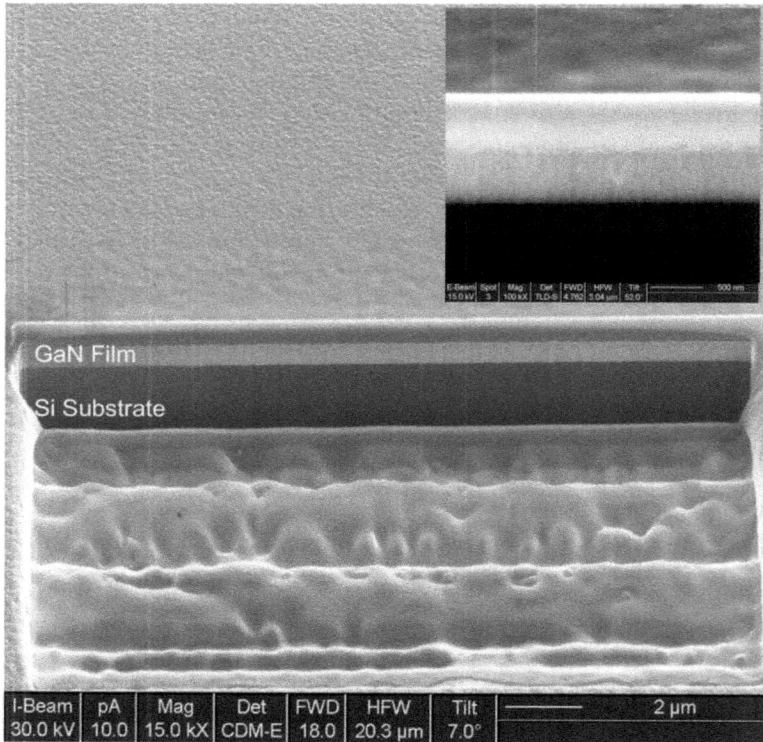

Figure 1.2. Ion-milled cross-section of polycrystalline GaN film deposited on Si wafer. The FIB has been used to remove the material in the foreground, revealing the film thickness and texture. There is also a protective strap of ion beam deposited platinum directly above the sectioned area. The section is 20 μm wide, 3 μm deep and the film is approximately 500 nm thick with individual GaN grains of less than 50 nm. The image is an ion beam induced secondary electron image. The inset shows a higher magnification secondary electron image (from the primary electron beam).

The simplest measurement we can perform is based on imaging, utilising secondary electron or secondary ions ejected from the sample by the energetic ion beam, where the highest spatial resolution can be below 1 nm. Focused ion beam can additionally be used to measure material properties beyond dimension, such as composition, texture and residual stress. Furthermore, we can construct or modify devices that can in turn also be used to measure dimension, magnetic or electrical properties. We can also construct artefacts with sufficient precision for them to be used to calibrate other equipment. Table 1.1 describes some of the areas we will discuss in later chapters. These topics will be divided into two broad categories; materials science applications targeted exclusively at understanding materials properties and composition, and FIB-fabricated devices and artefacts that may also have some materials science applications, but can be used in a far wider measurement context.

Table 1.1. Examples of areas where FIB can be used as a measurement tool, or used to fabricate a device to make a measurement, and a brief method summary of each.

Measurement	Method summary
Imaging	Either secondary electron or secondary ion. Ultimate resolution determined by type of source. Primarily for topographic imaging, but ion channelling can lead to contrast in polycrystalline materials.
EDX	FIB-polished sections or slices for electron beam excitation of x-ray signals, measuring percentage concentration of elements present.
EBSD	FIB-polished sections or slices for measurement of grain size and texture via electron beam.
3D reconstruction	Sequential FIB slices to produce 3D reconstructions of either dimension, grain size and texture or composition of samples.
TEM foil	FIB-prepared TEM foil for dimensional, compositional determination with better than atomic resolution.
Residual stress	FIB-milled holes in the μm range combined with image correlation to determine residual stress levels near the sample surface.
SIMS	Inclusion of a mass spectrometer allows material composition determination with few 10 s nm spatial resolution of sample surfaces.
Atom Probe	Succesive atomic layers are removed from FIB-prepared sharp tips.
AFM tip	FIB-sharpened tips can achieve atomic resolution.
Dimensional artefacts	FIB-fabricated step edges and lines for calibration artefacts.
SQUIDs and Hall bars	FIB-fabricated magnetometers.

References

[1] www.bipm.org
[2] Kimothi S K 2002 *The Uncertainty of Measurements: Physical and Chemical Metrology: Impact and Analysis* (ASQ Quality Press)
[3] www.npl.co.uk/publications/guides
[4] Goldstein J, Newbury D E, Joy D C, Lyman C E, Echlin P, Lifshin E, Sawyer L and Michael J R 2007 *Scanning Electron Microscopy and X-ray Microanalysis* 3rd edn (Berlin: Springer)

Chapter 2

Focused ion beam

2.1 Introduction to the FIB instrument

It has been known for more than 100 years that charged particles can be controlled (accelerated and deflected) by electric and magnetic fields, and from these early fundamental experiments an understanding of charged particle optics has developed [1], leading to the first electron microscope [2] and more recently focused ion beam microscopy, with the first FIB instruments appearing in 1974/5 [3, 4]. From these early experimental FIB instruments has grown a series of commercially available tools that both compete with, and complement, the field of electron microscopy. These tools offer a range of techniques, from simple imaging to complex fabrication, that encompasses a rapidly growing field applicable to samples as diverse as biological cells and engineering alloys. The early experimental machines utilised many different technologies and ion species, but the modern instrument has evolved into three distinct families, albeit with many similarities and common features; gas field-ion source, liquid metal (or alloy) ion source and plasma ion source.

Figure 2.1 shows a state-of-the-art (as of 2015) plasma FIB system offered by the manufacturer, FEI Company, and exhibits many of the features common to all FIB instruments, irrespective of the ion source. All systems will feature a sample chamber, maintained under high vacuum; an ion column consisting of an ion source; accelerating, focusing and scanning optics; a series of detectors, at least for secondary electron and secondary ion detection; a stage with at least three degrees of freedom (but usually five, x, y, z tilt and rotate); and an interface to control the operation of the instrument and maintain basic system functions such as vacuum management. Beyond these common basic features many other options exist, such as gas injection, additional signal detection methods, tools for TEM sample removal, charge neutralisation and specialised mechanical test stages, to name a few (this is discussed in more detail in sections 2.3–2.5).

In addition to the ion column, a large number of FIB instruments will also incorporate an electron column. In most cases, the electron column, being the more massive, will be mounted vertically on the top of the chamber with the ion column

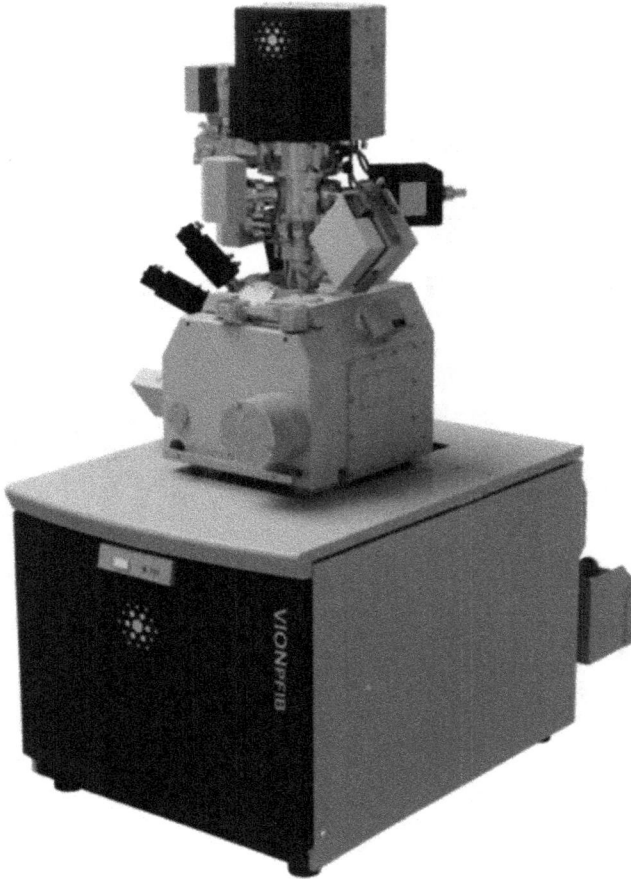

Figure 2.1. Photograph of a plasma FIB. The model is a VION and is supplied by FEI Company. Reproduced with permission from FEI.

side mounted at an angle determined by the geometry of the pole pieces of the two columns. The two columns will have one point in the chamber space where they both focus to the same spot (the coincident point), enabling the two columns to be used simultaneously, for example, the electron column monitoring a sample that the ion column is sputtering. Additionally, on two column instruments the stage will offer eucentric tilt, enabling the sample to be tilted between normal incidence for the electron and ion columns without the sample moving from the field of view of either column. The main purpose of an additional electron column is to provide non-destructive imaging of samples as the ion beam will cause sample damage even at very low ion doses. However, the electron column also enables the use of backscatter electron detectors, x-ray detectors and electron backscatter cameras, all characterisation techniques that can can be combined with FIB to either assist in determining where the ion beam should be focused or to characterise a material after ion beam exposure of internal features.

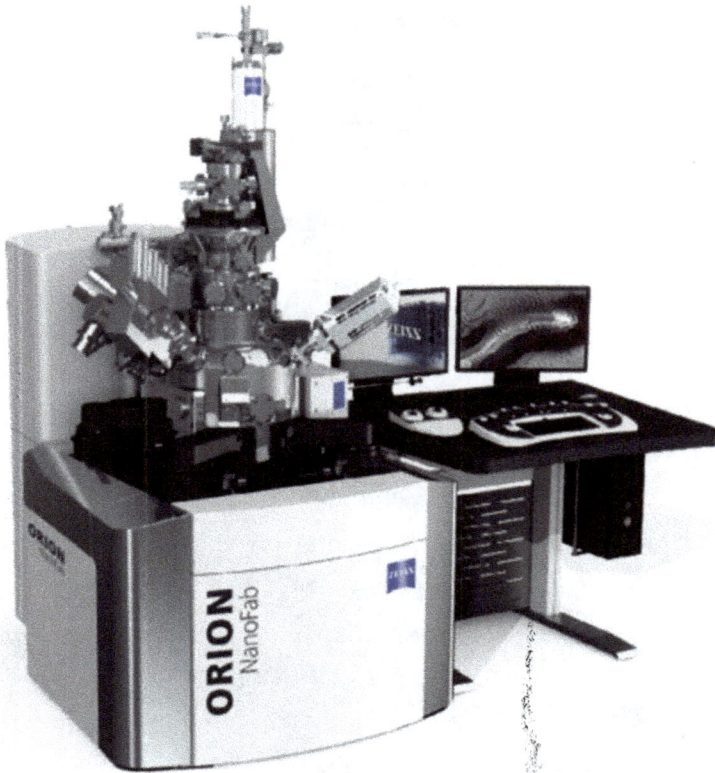

Figure 2.2. Photograph of a two column instrument. The vertical column is a He/Ne ion column and the column mounted on the left-hand side at an angle is a Ga ion column. The model is an Orion Nanofab and is supplied by Zeiss. Reproduced with permission from Zeiss.

One manufacturer (Zeiss) currently produces a two column instrument where both columns are ion columns (figure 2.2). On this instrument the vertically mounted column is a gas field-ion He/Ne column, used primarily for imaging, but the Ne ion is massive enough to achieve a measurable sputter yield. The second, side-mounted, column has a Ga ion source and is the main sputtering species as Ga offers substantially higher sputter yields than Ne (see chapter 3, section 3.3). This combination offers incredibly high-resolution imaging from the He, and two options for milling where the Ga can be used at high currents (nA) to rough cut structures and the Ne can then be used for a final polishing step at hundreds of pA or lower.

At the time of writing, FIB instruments are offered by the following manufacturers: FEI Company (USA and the Netherlands), Zeiss (Germany), Raith (Germany) Hitachi (Japan), JEOL (Japan) and TESCAN (Czech Republic), with ion columns also available from Orsay (France) and Ionoptika (UK).

2.2 Types of instrument

The main differences present in the three types of FIB column are almost all contained in the physics and chemistry of the ion source. Beyond the source, all columns are

Suppressor & LMIS

Extractor Cap

Beam Acceptance Aperture (BAA)

Lens 1

Beam Defining Aperture (BDA)
Quadrupole

Beam Blanking Plates

Blanking Aperture

Deflection Octupole

Lens 2

Sample

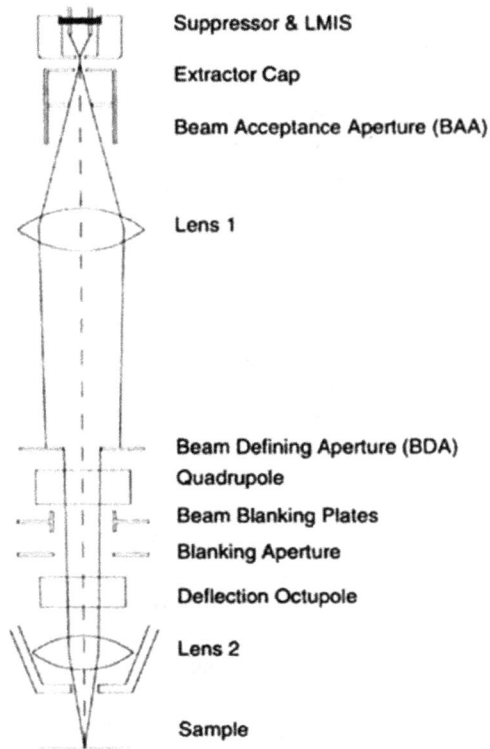

Figure 2.3. Schematic diagram of an FIB column equipped with a Ga liquid metal ion source. Reproduced with permission from FEI Company.

essentially very similar, largely using electrostatic lenses and beam deflectors and simple physical apertures to control, shape and scan the beam. Figure 2.3 shows a schematic diagram of a Ga ion column. At the very top of the column is the ion source, the extractor governing the accelerating voltage and a beam acceptance aperture that controls the maximum ion current that can flow down the column. Below this point is a series of lenses, which shape the beam, an interchangeable aperture of different hole sizes that determines final beam current, and a beam blanking mechanism to prevent the beam exposing the sample during the beam write back to the start of new image frames or milling pattern repeats. Finally, a second octupole lens deflects the beam for scanning and a final lens forms the focused beam of ions on the sample. Not shown is the vacuum system, usually composed of ion getter pumps, which keeps the whole column at vacuum levels lower than at most 1×10^{-7} mbar, and a column vacuum isolation valve that avoids the need to vent the column when opening the specimen chamber to air.

Table 2.1 lists the commonly available ion species and their source type, and gives an overview of the general properties that can be expected from sources of each species. This list is not exhaustive and other species are available, but at the time of writing others are not generally in use, except in specially constructed instruments.

Table 2.1. Commonly available focused ion beam sources.

Source	Type	General properties
He	Gas field-ionisation	Primarily used for imaging, almost zero sputter yield, very high resolution, very low beam currents (few pA), limited stability.
Ne	Gas field-ionisation	Alternative to He, limited sputter yield but high resolution for imaging, low beam currents (pA), limited stability.
Ga	Liquid metal	Most common source, modest beam currents (up to a few 10 s nA), relatively high mass allows reasonably high sputter rates, good resolution—down to few nm, good lifetime and stability.
Si	Liquid alloy	Available as an alloy component only (with Au), column requires Wein filter, current limited to hundreds pA, commercially available but still limited use at present.
Au	Liquid alloy	Alloy component with Si (possible alternative is Ge), offers high sputter yield compared to low value of its alloy partner Si, other properties as Si.
Xe	Plasma	Very high currents possible (few μA) and high mass allows large volume removal up to three orders of magnitude higher than Ga, resolution approximately an order of magnitude poorer than Ga, Other noble gases available (Ar, Kr) but not particularly common.

An example of a non-standard FIB is an H_2 instrument used for extreme ultraviolet (EUV) mask edit and repair. This instrument is currently manufactured by Hitachi High Technologies, but owing to the high development cost and highly specific use only a small number will be constructed.

2.2.1 Gas field-ion source

The gas field-ion sources used are predominantly those with the lowest mass species, the noble gases He and Ne, although heavier noble gas species can also be used. Sources of this type have been reported as far back as the earliest FIB instruments [3–5] and are derived from very similar sources invented in the 1950's [6] and subsequently used for the surface science technique, field ion microscopy. In the modern form of this source a tungsten tip is sharpened so that at the very end of the it only a trimer of atoms is present. This atomically sharp tip is placed in close proximity to the extractor, which is held at very high potential with respect to the tip. Gaseous He or Ne is kept at low pressure in the ion gun and when the molecular gas enters the extremely high electric field between the tip and extractor the gas is ionised and the extractor accelerates the ion into the column (figure 2.4). The emission current density is exponentially dependent on the electric field strength, and owing to the very close proximity of the extractor and extremely sharp tip, the increase in electric field at the tip causes a large increase in current. Just a few nanometers away from the tip the electric field lines seen by the ions appear almost parallel and the

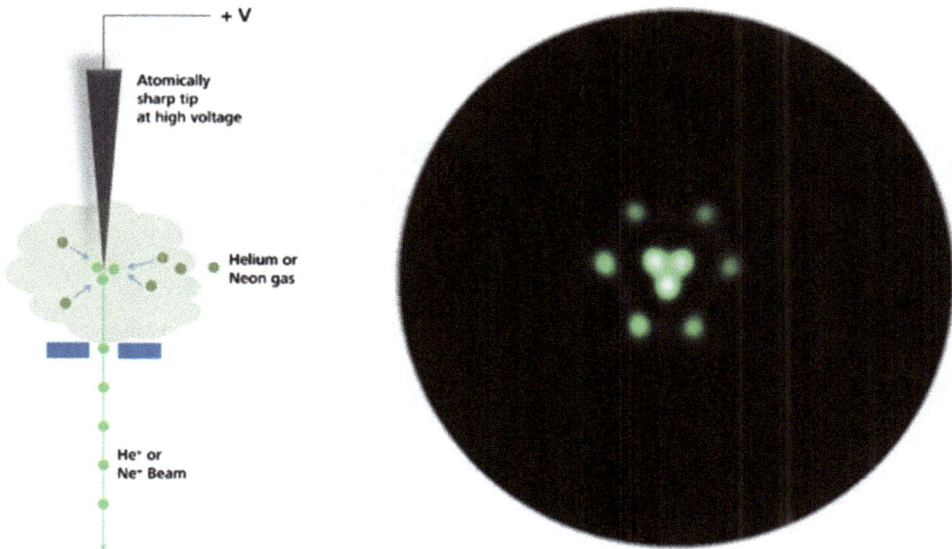

Figure 2.4. Schematic diagram of a gas field ion source and an image of the source tip showing a trimer arrangement of tip atoms. Reproduced with permission from Zeiss.

ions are immediately accelerated into a narrow, near-collimated beam. This leads to an ion source of extraordinary brightness so long as the tip is atomically sharp.

As the tip requirement is so demanding in this type of source, any change in the geometry of the tip, even on the atomic scale, leads to a loss of brightness. Tips do tend to diminish over relatively short timescales and must be regenerated frequently and replaced periodically. For this reason, sources of this type can be considered as having limited stability when compared to liquid metal ion sources. Additionally, even though the source can be considered to be very bright (i.e. high ion current in a very small beam spot) and is therefore excellent for imaging, it supplies only a modest maximum beam current of less than 100 pA, and if the low mass ion species used in this type of source is also considered, it is not particularly useful for sputtering sample volumes other than for very fine detail and final polishing of surfaces.

2.2.2 Liquid metal ion source

By far the most common ion source used in FIB instruments is the Ga liquid metal ion source (LMIS). Offering beam currents from as little as 1 pA to more than 50 nA, very stable operation and lifetimes in excess of 1500 h for continuous use, the Ga LMIS has dominated sales of commercial FIB instruments from their first availability to the present day. The main reasons for the dominance of Ga as an ion species is that is has both a low vapour pressure and a low melting point, it is relatively heavy and so offers a high sputter rate, and furthermore it is not particularity reactive. Also, the huge range over which Ga remains molten before it boils allows for a large safety margin during operation and enables additional source heating to move Ga from the reservoir to the extraction tip.

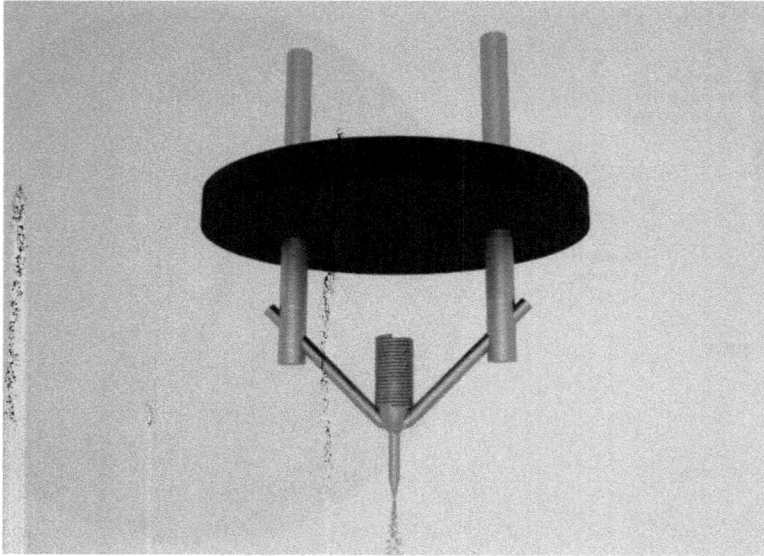

Figure 2.5. Computer model of an LMIS.

Figure 2.5 shows a computer generated model of a typical Ga ion source. It is a very simple device with some similarity to a thermionic electron source from an SEM, and is approximately 1.5 cm in diameter. The two posts fit into a socket and are electrically connected to a low voltage but relatively high current power supply. The applied current ohmically heats the thinner tungsten wire, causing the Ga contained in the coil reservoir to melt. By carefully controlling the heating of the source, Ga then wets the wire and will flow to the tip. As with the gas field-ion source, successful operation of the source is determined largely by the tip geometry. However, in the case of the LMIS, and somewhat counter-intuitively, the tip is actually quite blunt. The key to the operation of the LMIS is the formation of a Taylor cone [7]. As is shown in figure 2.3, very close to the tip and at a high negative potential with respect to the tip, is the extractor. The electric field from the extractor will attempt to pull ions from the tip, but the surface tension of the liquid Ga will attempt to repulse this. As these two forces are balanced, a cone shape begins to form with convex sides and a rounded tip with a tip radius of only a few nm. Once the extractor voltage is increased slightly, a stream of Ga ions will be emitted from the end of the cone and accelerated into the column. The maximum ion current that such a tip can emit is in the low μA range, although typically a maximum of a few tens of nAs is allowed through the acceptance aperture and down the column.

The large range of beam current that the Ga LMIS offers is a very good compromise in terms of imaging and sputtering. On current state-of-the art Ga ion columns an imaging resolution of better than 5 nm is easily achievable in the pA beam current range, while at higher currents in the tens of nA sputtering volumes of greater than 1×10^6 μm are achievable in a few hours in most materials. The simplicity, reliability and long-demonstrated usefulness of Ga LMIS sources suggests that they will continue to be the dominant ion source for the foreseeable future.

2.2.3 Liquid metal alloy ion source

As the name suggests, liquid metal alloy ion sources (LMAIS) are very similar to LMIS, and the basic operation of the source is indeed similar. The principal difference is that the Ga metal is replaced with a low melting point alloy, usually of two components. The two most common alloys are Au/Si and Au/Ge, although others exist and in principle the number could be expanded very considerably to include ternary and more complex alloys with suitable research efforts.

Turning to Au/Si, this binary alloy forms a low melting point eutectic at 536 K with the addition of 2.3% (weight) Si to Au. Due to the large difference in atomic mass this equates to 18% atom per cent Si in the alloy. Assuming that the source also emits in this same proportion, as the vapour pressures of the two are very similar, then almost one in five of the emitted ions will be Si. The simultaneous emission of two ions with such massively different atomic numbers (Si 14, Au 79) in a conventional ion column will, however, lead to an almost unusable instrument. As the two species have very large mass differences, this will lead to very dissimilar deflection angles due to the forces applied by the lenses and scanning octupoles in the column for each species. Quite simply, it would only be possible to focus and scan one species with the other chronically out of focus but still irradiating the specimen.

Fortunately, this problem was solved (albeit unknowingly) at the end of the 19th century [8] with the invention of the Wien filter, a device that was originally used in the study of anode rays, but that can also be used in mass spectrometry. Figure 2.6 shows a schematic diagram of a Wien filter. The device works by applying an orthogonal electric and magnetic field that the ion passes through. A charged particle in an electric field will experience a force proportional to its charge and the field strength. Similarly, the same particle in a magnetic field will experience a force dependent on its velocity as well as its charge and the field strength. As the force on the particle due to velocity is only affected by the magnetic field, it then becomes possible to balance the electric and magnetic fields so that no overall force is applied due to charge. By then increasing or diminishing the electric and magnetic field strengths together while keeping them balanced, it becomes possible to deflect particles that have different velocities. In a Au/Si LMAIS the ions will have significantly different velocities due to their large mass difference as they are accelerated from the source by the extractor potential. The Wien filter can then easily select ions of one species or the other by changing the Wien filter field strengths. In an LMAIS column (often referred to as an $E \times B$ column) the Wien filter filters the two species before the beam aperture and final lens and scanning system.

While it is incredibly useful to have a FIB system with interchangeable ion species, where large sputter yields are obtainable with the high mass Au ions and fine polishing or less-destructive imaging is possible with the lower mass Si, there is a limit to the maximum current that these sources can deliver. Furthermore, at high beam currents a large number of one species must be being deflected in the Wien

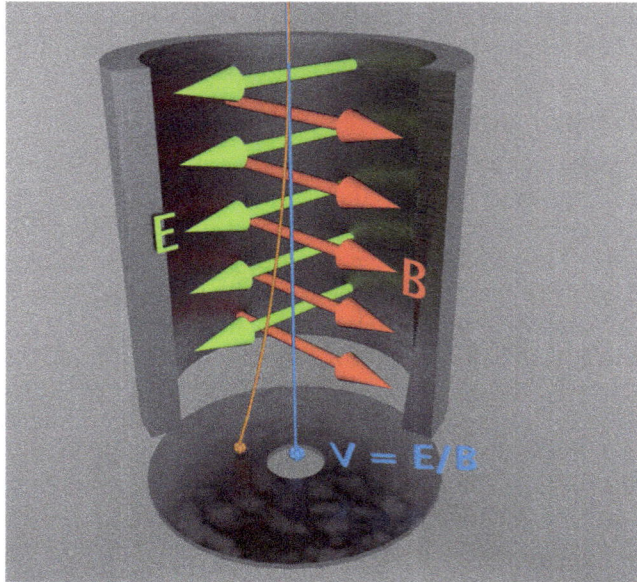

Figure 2.6. Schematic diagram of a Wien filter. Electric (green) and magnetic (red) fields are aligned orthogonally and balanced so that as a particle enters the filter it will be not be deflected if it has the correct velocity (blue particle) and exits the filter. Whereas if it has a different velocity, due to different mass at the time it was accelerated, it will be deflected (orange particle) and not exit the filter.

filter and hence there is a high probability of collisions and therefore scattering with the second species rendering the filter less effective. When one also considers that the source inherently delivers a fraction of the current as one species or the other, and then one of those species is filtered out, the typical maximum operating currents that can be used are in the few hundreds of pA. This limitation should not, however, detract from the usefulness of such sources, for example where Ga should be avoided as it may alloy with the sample, or where it is deemed too aggressive and hence the lighter Si may be a desirable option.

2.2.4 Inductively coupled plasma ion source

The final and most recent addition to commercial FIB sources is the plasma ion source [9]. Using noble gases, principally Xe, although it is also viable with Ar and Kr, it offers by far the highest deliverable beam currents of any type of commercial FIB system, with up to 5 µA of beam current available at the sample, while still offering sub-30 nm imaging resolution at the lowest currents, in the low tens pA range. As the ion species is supplied from a simple gas cylinder and regulator, and is of such simple construction, the lifetimes of these sources is both long and stable. Indeed, providing the gas supply is maintained they can be run continuously, potentially for years, a situation where a LMIS would exhaust the ion species in less than three months owing to the small amount of available material. Figure 2.7 shows a simplified cutaway schematic model of an inductively coupled plasma source.

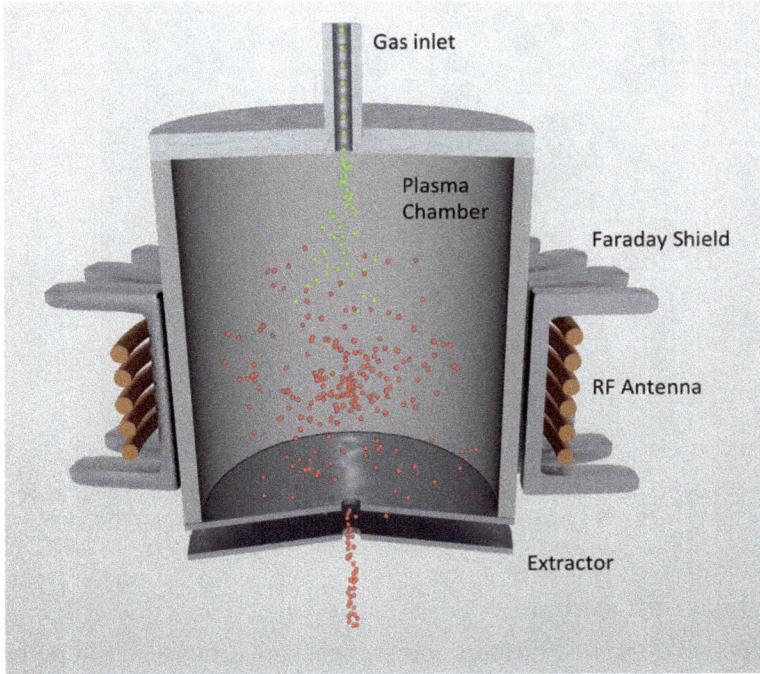

Figure 2.7. Schematic cutaway model of a simple inductively coupled plasma source. Gas is introduced at the top of the source (green), where it fills the dielectric plasma chamber at low pressure. An RF antenna ionises the gas (red), where it is extracted by the high voltage extractor and enters the column. The Faraday shield minimises possible capacitive coupling from the antenna to the plasma.

The main advantage the plasma source offers is the extremely large volumes (by FIB standards) that these instruments can sputter ($>100~\mu m^3~s^{-1}$). The combination of the high beam current and mass of the Xe ion means the plasma FIB can produce structures that are at least one order of magnitude larger in all directions than is feasible with a Ga FIB in an equivalent time. For example, figure 2.8 shows a cube of Nb of 360 μm × 360 μm that was fabricated by Xe ion sputtering in under 2 h. This structure is close to the size that would be visible to the naked eye and would have taken in excess of two days to mill with a Ga ion beam. Furthermore, the range of a 30 kV Xe ion in Nb is considerably shorter than that of Ga and so the damaged zone will also be considerably reduced at the sample surface. One further consideration with the plasma FIB source is that similarly to the gas field-ion source it operates with noble gases. In some cases, having an ion species that is incapable of reacting with or alloying with the specimen is highly desirable, for example if one were to mill some of the semiconductor compounds, such as GaAs, not adding extra Ga from the ion source would eliminate one possible source of measurement error. These sources, however, are not particularly bright when compared to LMIS sources, which are two orders of magnitude brighter, and they could be considered quite poor when compared to gas field ion sources, which are some five orders of magnitude brighter.

Figure 2.8. A niobium cube sample (360 μm × 360 μm), created for dynamic compression testing and fabricated in only 2 h with a Xe plasma FIB. Reproduced with permission from FEI Company.

2.3 Gas injection systems

The modification of sample surfaces and the removal of material by sputtering is by far the greatest use of all but the lightest element FIB instruments. However, not all specimens will mill at a constant rate, some are more prone to redeposit back on the sample more than others, and when milling deep holes the hole can become too deep to remove material from the bottom effectively, preventing any further increase in depth. To alleviate some of these problems in many materials, a range of chemicals was developed to assist sputtering by introducing a reactive component. One of the most common of these, and also one of the simplest, is water, which can be used to enhance milling in carbon-based materials. The use of very low pressure water vapour in the chamber assists with milling materials as diverse as polymers, such as PMMA, or diamond, increasing milling rates by an order of magnitude or more. The water vapour is introduced by a thin needle tube in close proximity to the sample surface. The water contained in a moderately heated and evacuated crucible sublimes from magnesium sulphate heptahydrate (commonly known as Epsom salts) and is introduced into the sample region on opening a valve connecting the crucible to the chamber. The combination of ion sputtering and the reactive component enables high milling rates and significantly reduces redeposition. Other enhanced etching chemistries are iodine and xenon difluoride, used primarily in the semi-conductor industry.

Using essentially the same components, but with different, less reactive chemistry, another use of GIS is the deposition of materials. The single most common deposition material is Pt, while W is also widely used. In both cases (and several others), these are organometallic compounds, where they either boil or directly sublime at modest temperatures of less than 100 °C. In the same method as the etching gases, a valve allows the gas to travel from a crucible down a needle tube to the sample surface, where

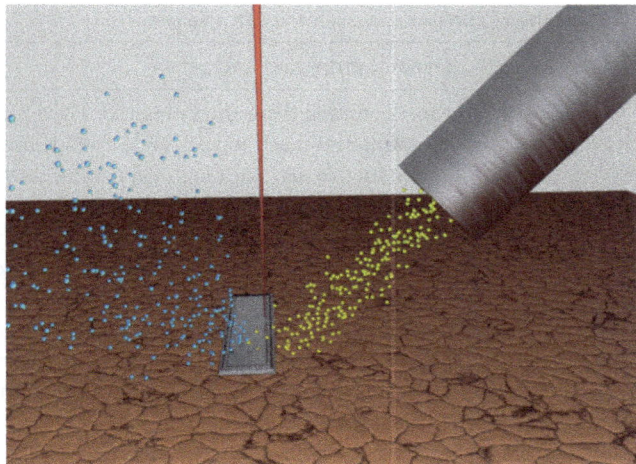

Figure 2.9. Schematic digram of a GIS used for deposition. The deposition gas (yellow particles) is introduced into the chamber in close proximity to the sample surface. The ion beam (red) is energetic enough to crack the gas molecule, depositing the desired species on the sample surface (grey). The volatile species (blue) from the gas is not deposited and is pumped away by the vacuum system.

it initially adsorbs and is then decomposed into volatile and non-volatile components by the ion beam, shown schematically in figure 2.9. In the case of both Pt and W in a Ga ion beam FIB the deposit is largely composed of a mix of the metal from the compound, Ga from the ion beam, with the balance being some of the carbon, also from the compound. The exact chemistry of the deposit varies with beam conditions and gas chemistry, but the properties of these deposits is to a first approximation metal-like, in that they are fairly conductive (actually superconducting at 4 K in the case of W), fairly hard and shine like a polished metal surface when viewed with an optical microscope. The uses of the deposition gases largely fall into three categories. Firstly, they are used to deposit protective straps that preserve the leading edges of ion-milled faces and TEM foils from the beam tails of the ion beam (and sometimes also give good contrast variation for imaging). Secondly, they are used to attach things together, for example TEM lamella lifted out and attached to TEM grids. Thirdly, they are used to create and repair electrical leads in electronic devices (and also masks used to create devices by optical lithography). For some of these activities, several of the available chemistries can be used, for example for sample edge retention Pt, W and C are interchangeable and just have different deposition rates, but other uses require specific chemistries, for example the use of tetraethyl orthosilicate as an insulating layer in circuit edit applications.

Table 2.2 lists the most commonly available gas chemistries that are supplied with FIB systems. This list is not exhaustive and several others are available, including several other metals and enhanced etches, but in most cases these are somewhat experimental in nature and tend not to be offered by manufactures as original equipment. A more complete list is given in chapter 3 of [10].

Table 2.2. Commonly available gas chemistries available in FIB systems.

Common name	Chemisty	General properties and uses
Platinum	$C_5H_4CH_3Pt$ $(CH_3)_3$	Most common deposition material, Used for protective caps, circuit and mask repair, TEM foil lift out. Reasonably conductive and with high deposition rates.
Tungsten	$W(CO)_6$	Alternative to Pt, lower deposition rate, slightly better conductivity, used a lot in circuit edit applications.
Carbon	$C_{10}H_8$	Another alternative to Pt, but poorer conductivity, low sputter yield makes it good material for protective caps, also shows as low contrast on high contrast substrates.
Insulator	$Si(OC_2H_5)_4$	Used primarily in circuit edit and repair, forms near-perfect SiO_2 if deposited by electron beam, alternatives are siloxane-based chemistries.
Gold	$C_7H_7O_2F_6Au$	Expensive chemical almost exclusively used with biological materials.
Carbon mill	$MgSO_4 \cdot 7H_2O$	Uses the contained water content only as a reactive component in milling high carbon content substrates. Particularly effective with polymers.
Fluorine	XeF_2	Enhanced milling of silicon and silicon nitride and some metals. Is reactive enough to use with electron beam in some cases.
Iodine	I_2	Enhanced milling of silicon and particularly effective with GaAs and InP.

2.4 Patterning options

One of the key features of all FIB systems is the ability to steer the beam around the sample and produce the desired ion-milled structures. In essentially the same way that the combination of scan coils and beam blanker scan the beam to produce an image, so these same components can be used to produce far more complex scan patterns than the simple rectangles used in imaging. Essentially, every milling pattern requires a combination of beam current, beam dwell (time the beam sits at one point), the beam step size in the x and y directions, the beam scan geometry and the number of times to repeat the geometry (figure 2.10). Taking these in turn, the beam current will determine the amount of material sputtered in a given time. The beam dwell will perform a similar role, but in general long dwell times (>10 μs) tend to redeposit more material than shorter times. The beam step size for ion sputtering is in most cases set so that adjacent beam spots overlap by about 50%, but there will be some small rounding error associated with these steps as most FIB systems will use 16 bit digital to analogue conversion (DAC) to convert digital computer output to analogue scan coil voltages. The geometry is the shape of the milled region and the number of repeats is combined with current and dwell to mill the sample to the correct depth. It should be stated that every material will have a unique sputter rate, often undetermined, and to some extent the setting of these parameters can be a case of trial and error.

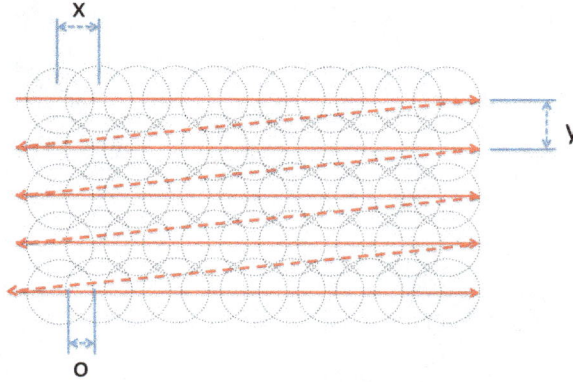

Figure 2.10. Schematic diagram of the ion milling process. The dashed circles denote the beam spot on the sample surface. The beam is stepped from one position to the next along the solid red line in equal steps, where 'x' denotes the step size along the line and 'o' denotes the overlap of beam spots, usually expressed as a percentage of the beam spot. At the end of the line the beam is blanked and swept back to the start of the next line (dashed red line). The line spacing 'y' is usually equal to the step size in 'x', although it need not be. An alternative is to move the beam in a serpentine path, where it continues on the next line in the opposite direction to the preceding line.

All standard commercial FIB systems will offer at least a basic patterning capability, including simple geometric entities such as rectangles, circles and lines along with a few specially designed milling patterns for cross-sections and sample polishing, along with automated procedures for TEM foil preparation and sequential slice milling. Additionally, most will offer the ability to interpret greyscale bitmap images where the grey level at each pixel in the image is used to set the dwell time of an equivalent pixel in the milled pattern. An example of such a pattern is shown in figure 2.11, where the contrast in the image is derived from a combination of ion milled depth and implantation damage in Si.

While the standard pattern capability is adequate for most FIB applications, for complex fabrication some manufacturer extensions and third-party systems are currently available that add considerable sophistication to patterning capabilities. These systems all adopt similar interfaces to electron beam lithography (EBL) systems, to a lesser or greater extent. NPVE is a set of tools from Zeiss that extends the patterning capability of their FIB systems and handles multiple patterning entities and the option to set individual co-ordinate points for each beam dwell point (as well as offering advanced SEM imaging tools in the same package). A considerable step closer to an EBL-like system is GDStoDB, an extension to FEI's patterning capability that enables the import of CAD drawing files. Finally, two systems that are in fact EBL systems but are also capable of controlling ions columns are provided by Raith [11] and NPGS [12]. Both of these systems offer CAD drawing input, use of multiple entity types and many drawing layers within each CAD file. Systems such as these are capable of turning FIB into very sophisticated fabrication tools, including, as has been recently demonstrated, curved milled surfaces of very precise geometry [13].

Figure 2.11. Secondary electron image (from primary ion beam) of an ion-milled greyscale bitmap image of the Moon in Si.

2.5 Other equipment and techniques found on FIB instruments

While not part of a standard FIB instrument, and in most cases dependent upon an electron column being present, it is worth briefly mentioning other, often third-party equipment that can be found on FIBs, and their intended uses, some of which are described in more detail in chapter 4. For example, FIB systems that are predominantly used for materials characterisation will often be found equipped with energy or wavelength dependent x-ray detection (EDS, WDS) [14]. Such systems are used where the chemical composition of the specimen is required. Characteristic x-rays are not generated in ion cascades and so such systems rely on an electron column also being fitted to the instrument. In cases where these systems are found, the FIB will often have sputtered the sample, exposing the internal geometry that will then be imaged by the electron beam. By using FIB to expose below-surface details and combining this with SEM and EDS it is possible to sample the chemistry of the specimen considerably more deeply than with SEM and EDS alone. Another electron beam dependent technique is electron backscatter diffraction (EBSD) [15], a technique to determine the crystallography of samples. As with EDS, the electron beam is the sampling tool, but once more the FIB can be used to expose facets within the sample.

Until fairly recently, FIB has not been used significantly with biological samples. The limitations imposed on samples of this type by the vacuum and the destructive nature of the ion beam (and often even the SEM beam) to delicate tissue and other biological samples has precluded their widespread use. Furthermore, the fact that many biological samples for TEM studies can be prepared using well-established methods such as microtome has further reduced the uptake of FIB in this area. However, the addition of a cryogenic stage to a FIB enables its use with biological samples [16]. By freezing the sample quickly and keeping it cold using pumped nitrogen gas cooled from liquid nitrogen it is possible to study samples without the need for chemical fixing and the artefacts that may introduce. Furthermore, the highly accurate site-specific nature of FIB sectioning means that the sample can be sectioned with much higher precision than with a microtome. Cryogenic stages have also found use with metallic and other samples, such as the III–V compounds semiconductors (for example GaAs and InSb), where the low temperatures help to reduce formation of low melting point eutectic phases during sectioning with Ga ion beams.

One addition that is also found on some FIB systems is a means of charge neutralisation. As the sample is being irradiated with positively charged ions during milling, if the charge cannot escape, for example in ceramic or polymeric samples, the beam can be deflected to the extent that the ion-milled area will have the appearance of having moved during milling. In extreme cases, the beam deflection can be so significant that it is actually impossible to ion mill the specimen with any certainty and milled sections cannot be completed. One solution to this problem is to flood the area being milled with particles of the opposite charge, i.e. electrons. A low-energy electron flood can be directed at the sample surface and can deliver sufficient current to neutralise the charge from the ion beam even at the highest currents. It should be noted that on FIBs equipped with an SEM column the electron gun in the column is unlikely to deliver sufficient current to achieve the same effect if it is a field emission electron source, but it may just be enough in thermionic source columns with the beam defocused.

One key use of FIB is the production of TEM samples (lamellae), and in this area FIB excels, producing extremely high-quality and targeted thin sections of many samples that it would often be impossible to produce using other conventional methods. However, when one considers the sputtered volume a typical FIB can remove, particularly if it is not a plasma source instrument, the TEM lamella is likely to be of the order of 10 to 20 μm wide and probably at most a similar depth or less. There are essentially two methods of removing the lamella once cut from the sample and attaching it to a TEM grid. The first is *ex situ* removal using micro manipulators and an optical microscope. This method uses a sharpened glass needle and electro-static attraction to lift the lamella from the substrate, after which it is deposited on a carbon thin-film covered TEM grid. This method can be relatively time-consuming and requires a very high level of operator skill. A further limitation is that the lamella must be made to its final thinness, as no further thinning is possible once it is on the grid. The second method of TEM lamellae lift out is an *in situ* one that requires a sharpened (usually tungsten) tip that can be manoeuvred in the chamber using piezo actuators. These actuators are capable of sub-micron motion (often few nm) over many mm of range with three degrees of freedom. The tip is positioned so that it is in contact with the lamella and the two are connected together, usually using Pt GIS. The lamella is then removed from the substrate while being imaged in real time with the SEM, if present, or an ion beam at low current if not, and attached to a copper TEM grid, again using the GIS. Once attached, the tip is removed from the lamella using the FIB beam and then the foil can be further thinned to its final dimension on the grid. While this method requires the addition of extra hardware in the chamber, the ability to carry out the lift in real time whilst imaging has meant that it now dominates TEM lamella fabrication and such equipment is usually offered by the instrument manufacturers as a factory-supplied option.

References

[1] Oatley C W 1982 The early histrory of the scanning electron microscope *J. Appl. Phys.* **53** R1–R13

[2] Von Ardenne M 1937 Improvement in electron microscopes (British patent no 511204) convention date (Germany) 18 Feb 1937

[3] Levi-Setti R 1974 Proton scanning microscopy: Feasibility and promise *Scanning Electron Microsc.* **1** 125–34

[4] Orloff J and Swanson L 1975 Study of field-ionization source for microsprobe applications *J. Vac. Sci. Technol.* **12** 1209–13

[5] Escovitz W, Fox T and Levi-Setti R 1975 Scanning transmission ion microscopy with a field ionization source *Proc. Natl Acad. Sci. (USA)* **72** 1826

[6] Muller E W 1957 Study of atomic structure of metal surfaces in the field ion microscope *J. Appl. Phys.* **28** 1–6

[7] Taylor G 1964 Disintigration of water drops in an electric field *Proc. Royal Soc.* **280** 383–97

[8] Wien W 1898 Untersuchungen uder die electrische entladung in verdannten gasen *Ann. der Physik* **301** 440–52

[9] Smith N C, Skoczylas W P, Kellogg S M, Kinion D E, Tesch P P, Sutherland O, Aanesland A and Boswell R W 2006 High brightness inductively coupled plasma source for high current focused ion beam applications *J. Vac. Sci. Techol.* B **24** 2902–6

[10] Yao N (ed) 2007 *Focussed Ion Beam Systems* (Cambridge: Cambridge University Press)

[11] www.raith.com

[12] www.jcnabity.com

[13] Langridge M T, Cox D C, Stolojan V and Webb R P 2013 The fabrication of aspherical microlenses using focused ion-beam techniques *Micron* **57** 56–66

[14] Goldstein J, Newbury D E, Joy D C, Lyman C E, Echlin P, Lifshin E, Sawyer L and Michael J R 2007 *Scanning Electron Microscopy and X-ray Microanalysis* 3rd edn (Berlin: Springer)

[15] Schwartz A J, Kumar M, Adams B L and Field D P (ed) 2000 *Electron Backscatter Diffraction in Materials Science* (Berlin: Springer)

[16] Mulders H 2003 The use of a sem/fib dual beam applied to biological samples *GIT Imag. Microsc.* **5** 8–13

Chapter 3

Ion–solid interactions

3.1 Overview

The effects of energetic ions impacting on a sample surface are dependent upon the energy of the ions, the density of the sample and the angle of incidence of the ion impact with the sample surface. The dominance of Ga as the ion species in the vast majority of commercial FIB systems simplified this analysis considerably until very recently. However, as shown in the previous chapter, the number of available ion species in newer FIB systems, while not exhaustive, spans a large range of atomic masses and this mass difference can result in very different ion–sample behaviour.

Essentially, the effects we observe occur as a result of the impacting ions giving up their momentum in a series of elastic collisions with the sample atoms. The resulting cascade can lead to the displacement of hundreds of atoms within the target and if sufficient kinetic energy is transferred to surface atoms, greater than the surface binding energy of the material, these atoms will be ejected as sputtered particles. The impacting ion will in the meantime, depending on conditions, either implant into the sample or backscatter from its surface. When one considers that at, for example, the easily achievable, and in fact quite modest, beam current of 100 pA, greater than 6×10^{11} ions per second will be arriving at the sample surface, and that the beam will be focused into a spot of possibly 15 nm or smaller, it is clear that the local effects will be very considerable and easily observable. From the generation of secondary electrons and secondary ions for imaging, to the ejection of large numbers of both neutrals and ions from the sample during sputtering, it should be remembered that the collision of energetic ions with a sample is almost always going to be a destructive process and modifications, including unseen sub-surface damage, will be taking place as long as the sample is exposed to the ion beam.

Due to the nature of modern engineering materials such as complex alloys, multi-component thin films, single crystals such as Si wafer, polycrystalline materials, fully amorphous materials and compounds—a potentially infinite number of varieties—a complete treatment of the subject of ion–solid interactions is beyond the scope of this

doi:10.1088/978-1-6817-4084-3ch3   3-1

short book and the physics of ion–solid interactions will only be discussed in the last section of this chapter on modelling. For a much more extensive discussion on the subject of ion–solid interactons, the reader is directed to [1]; for a FIB-specific discussion see [2, 3]. This chapter aims to give the reader a basic understanding of the processes occurring in a FIB system and the limitations and advantages of various typical instrument parameters. Beginning with simple imaging, we move on to sputtering, ion range and damage, the effects of redeposition, alloying and heat generation. Finally, we introduce two very useful computer programs that use firstly, Monte Carlo simulation to model ion collisions with amorphous solids [4] and secondly, a numerical modelling program (SUSPRE) [5]. These are discussed in 3.4.

3.2 Imaging—secondary electrons and secondary ions

During the ion irradiation of a specimen the collision cascade will generate secondary electrons. The term 'secondary electron' is commonly used in SEM and refers to those electrons not originating from the primary beam in these instruments. However, strictly speaking, any electron liberated via ionising irradiation is considered to be a secondary electron. Secondary electrons are generally accepted to have energies lower than 50 eV, with a peak energy below 5 eV, the main reason for this being that the inelastic mean free path for electrons above 5 eV drops rapidly to a few nm above this energy and hence most are not likely to escape the sample [6, 7]. The expected yield of secondary electrons from a primary Ga ion beam that will be detected is of the order of one electron per ion [8].

Secondary electron imaging is the primary imaging mechanism used in FIB. Of course, a substantial number of FIB systems will have an electron column on the same instrument with this constituting the main imaging system. This should not, however, detract from the usefulness of secondary electron imaging from primary ions, and in the case of He (and Ne) ion machines this will still be the primary imaging mechanism. In gas field-ionisation He instruments the ultimate resolution approaches and indeed can beat that of the best scanning electron microscopes and the very small convergence angle of the beam also gives rise to excellent depth of field (see figure 3.1). The very high resolution is predominantly due to the way He interacts with the substrate. On impacting with the surface, all ions undertake a zig-zag path from one scattering event to the next until all the energy is given up to the substrate (or the ion escapes the sample). The heavier the ion, the greater, on average, the angles of deflection at each event tend to be, corresponding to the greater energy loss at each scattering event. The low mass of He results in a very large implant range compared to all other available FIB ion sources. Figure 3.2 shows the predicted ranges of the commonly available FIB ion species in two different substrate materials, Si and Au, as a function of energy. As can be seen, in a Si substrate the range is over 100 nm for He at only 10 kV, and in fact at 30 kV it is greater than 280 nm, whereas in Au at 30 kV the range is still almost 70 nm and longer than any other FIB ion species achieves even in the low density Si target. Additionally, with He ions there is an extremely low probability of sputtering and defect formation in the sample from atom displacements (meaning there are fewer additional cascade electrons from these displacements).

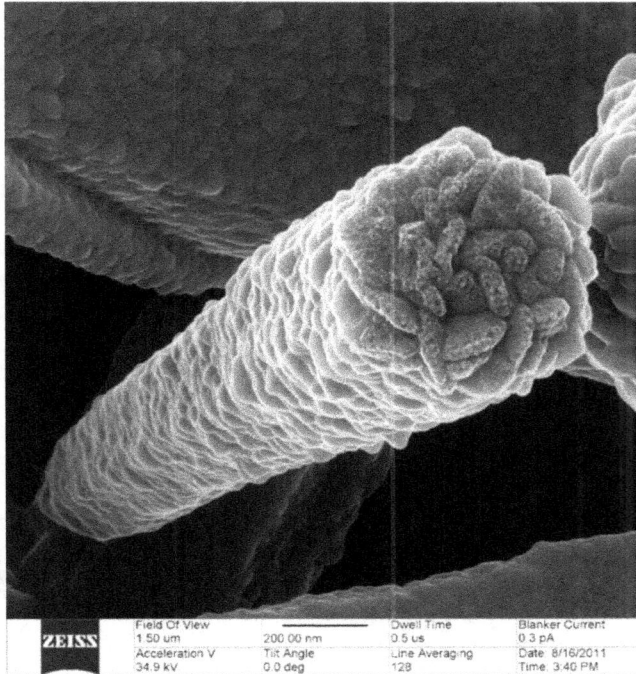

Figure 3.1. A secondary electron image of a palladium nanowire imaged in a Zeiss Orion He ion microscope. It demonstrates the outstanding resolution and depth of field that this instrument offers. Reproduced with permission from Carl Zeiss Microscopy, LLC.

The combination of long range, very low sputter yield and low defect generation results in a secondary electron emission area corresponding to not much larger than the beam spot size and hence sub nm imaging resolution. In comparison to He, the mass of all of the other available species of ion in FIB systems is large, the ion range therefore is much shorter and energy is given to the sample much nearer to the sample surface. The displacement of sample atoms near the surface also contributes to secondary electron generation in the cascade and the emission area will be larger than the primary beam spot size and the resolution poorer.

Although the heavier ions tend not to offer the highest imaging resolutions, secondary electron emission can still be a useful imaging process with these instruments. As with scanning electron beam imaging, scanning ion imaging is ultimately about obtaining contrast from the specimen. Secondary electrons generated from either an electron or an ion beam will generate contrast due to differences in secondary electron yield varying with local differences in average atomic number, crystallography and topography. With SEM secondary electron yield is shown to broadly increase with increase in average atomic number, whereas the reverse is true for scanning ion microscopy [9]. This difference in behaviour is due to two mechanisms. Firstly, in the case of scanning electron beams a significant part of the SE yield is generated from the backscattering electrons from the primary beam. As the sample

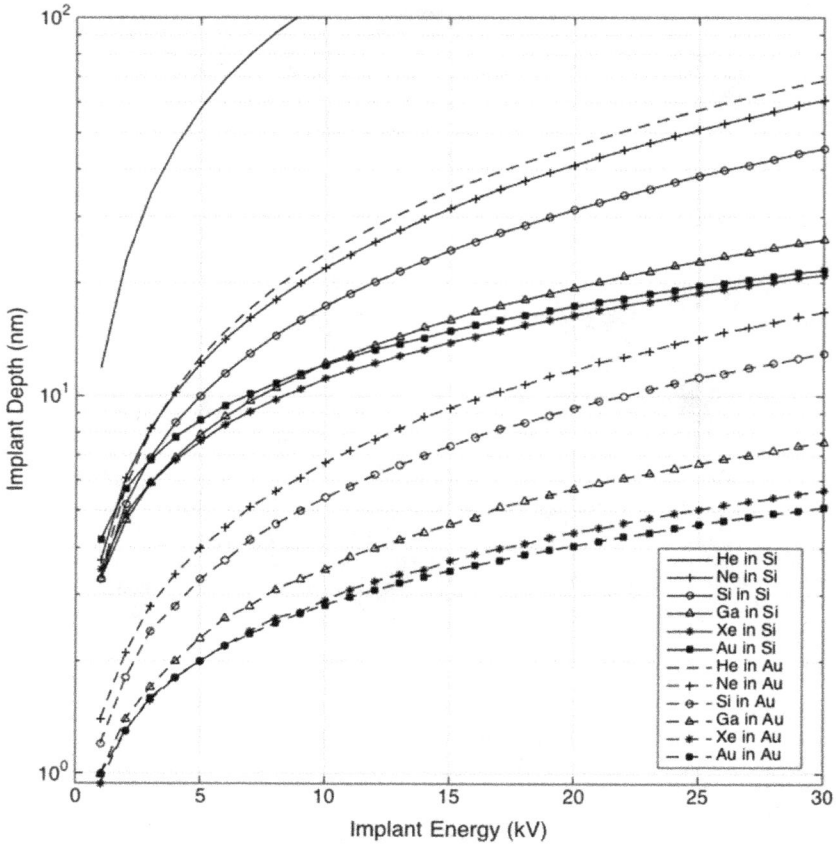

Figure 3.2. Ion ranges for typical FIB ion species as a function of implant energy in a low mass target (Si) and a high mass target (Au) impacting at normal incidence.

density increases, so does the backscatter yield and hence the portion of the SE yield from backscattering. In the case of scanning ion beam microscopy almost no backscattering of the ions occurs, and additionally, as the sample density increases, so does the stopping range of the ion, leading to lower probability of SE generation. This short ion range also leads to greater sensitivity to very thin films when imaging with the heavier FIB ions such as Ga, and often surface contamination is visible with 30 keV Ga ion imaging when compared to even sub 5 keV electrons.

By far the most striking effect of scanning ion imaging occurs when imaging polycrystalline samples and this is due to the channelling of ions along favourable lattice planes. In a polycrystalline sample some grains will be oriented so that ions can enter between lattice planes and will only undergo small angle scattering collisions, resulting in them travelling much further into the sample. This effect results in very low SE emission as the low angle scattering deposits a very small amount of energy at each collision. When the ion finally undergoes larger angle scattering deeper in the sample, the generated SE from the collision cascade will be

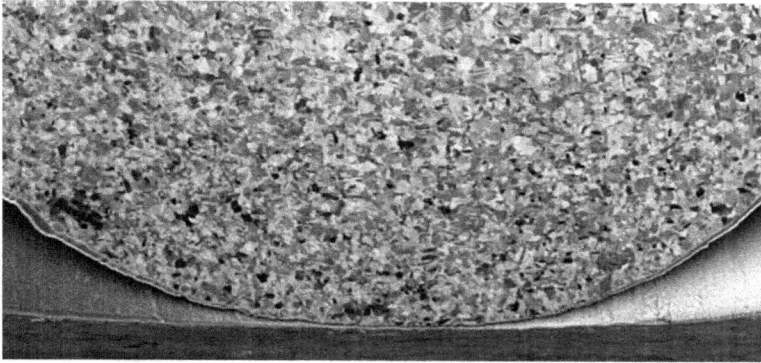

Figure 3.3. Secondary electron image generated from primary Xe ions of a polycrystalline metal sample (guitar string). The grain contrast is due to variation in secondary electron yield for different orientation grains from ion channeling. Reproduced with permission from FEI Company.

too deep to escape. By comparison, unfavourably oriented grains without open lattice planes will scatter incoming ions through much larger angles and generate substantial SEs in close proximity to the surface. An example of an SE image of a polycrystalline metal generated by Ga ions can be seen in figure 3.3.

While secondary electron yields are typically one electron per incident ion, sputtered atoms per ion typically range from less than one to a few tens. However, only a very small fraction (as low as one-thousandth) of the ejected substrate atoms will be ionised and can be collected for secondary ion (SI) imaging [8]. SI imaging does not find widespread use in the FIB community owing to inferior resolution compared to SE imaging, while additionally the low yield and inherently flat contrast of the images requires slow scan rates, which in turn translate to substantial ongoing sample modifications during continued scanning. However, on occasion complementary or otherwise inaccessible information may be obtained. For example, in samples with non-conductive regions the positive charge of the ion beam will suppress the emission of SEs due to charge build-up. These regions may appear very dark in SE images, but the charge will help to enhance SI emission and will appear brighter in SI images [3].

3.3 Ion milling—ion range, sputter yield and damage

By far the largest use of FIB instruments is in the removal of material by sputtering of samples with incident ions. While the He (and to some extent Ne) based machines are primarily used for imaging, all of the other species are predominantly used for the removal of sample material to study the internal composition, structure and morphology, or to shape the sample to perform a new function. The two key parameters to consider in this process are the removal rate of sample material, as this limits the volume we can access in a given time, and the resolution we can achieve when removing material, as this may limit the information we can obtain from the volume.

Figure 3.4 shows the calculated sputter yields for two different pure target materials, Si and Au, for the commonly available ion species over the normal range of accelerating potentials. As can be seen, the sputter yield for He is negligible for

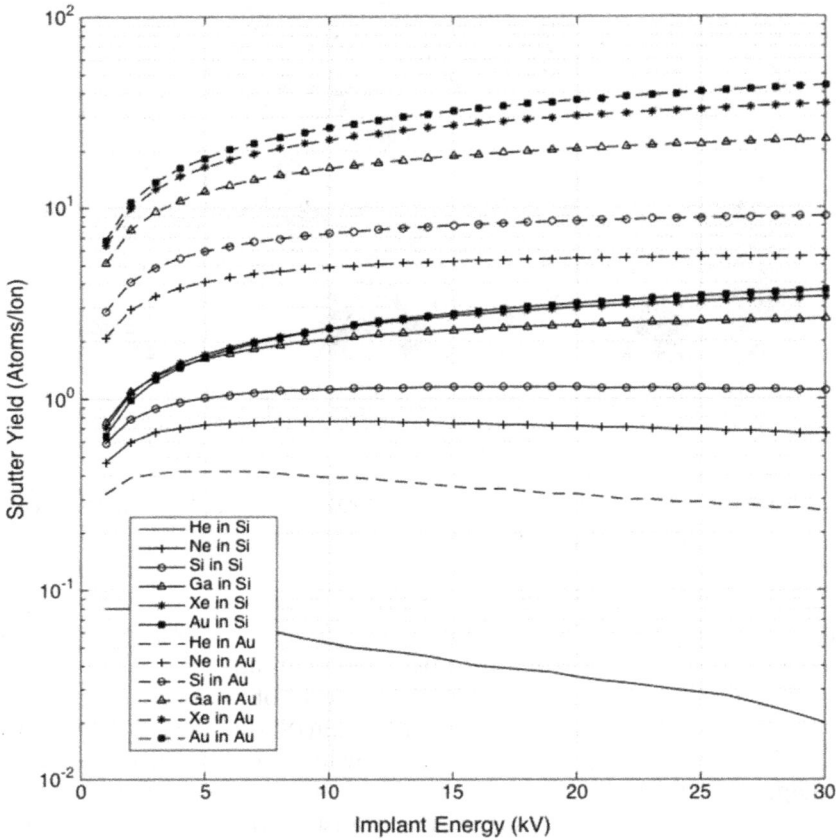

Figure 3.4. Expected sputter yields for a low mass target (Si) and a high mass target (Au) as a function of implant energy for selected FIB ion species impacting at normal incidence.

Table 3.1. Generalisation of expected range and sputter yield for different combinations of ion mass and target density.

Ion mass	Sample density	Sputter yield and range
Low	Low	Sputter yield is low and ion range is very high.
Low	High	Sputter yield is moderate and ion range is moderate.
High	Low	Sputter yield is moderate and ion range is moderate.
High	High	Sputter yield is very high and ion range is low.

both target materials, with all other ions showing higher sputter yield in the higher density targets, and as the ion species mass increases, sputter yield also increases. This behaviour is generalised in table 3.1. It should be stated that the models used to calculate these values assume that the material is fully dense, but that it is also amorphous (i.e. there is no ion channeling), and that it is pure. These limitations and their agreement with real materials are discussed in the last section of this chapter.

An interesting observation displayed in figure 3.4 is the weak dependence of sputter yield as ion accelerating voltage increases above around 5 keV for all the different ion species. This can be explained very simply, as although the incident ion has more energy available to sputter sample atoms as accelerating voltage increases, the higher accelerating voltage also increases the ion range, resulting in the available energy being deposited deeper in the sample, where it cannot be transferred to surface atoms overcoming their surface binding energy.

So far, the discussion of sputter yield and ion range has concentrated on incident ions impacting on the sample surface at normal incidence. In a very large number of cases this would be considered a standard mode of operation and, for example, when ion milling a cross-section or TEM foil the usual initial cut removes a block of material, exposing the face of interest where the material has been sputtered with normal incidence ions. However, when polishing the face of the cross-section the ions will be arriving at very high incidence angles close to 90° from normal. Figure 3.5 shows the effects of changing the incident angle of Ga ions on an Si target for three

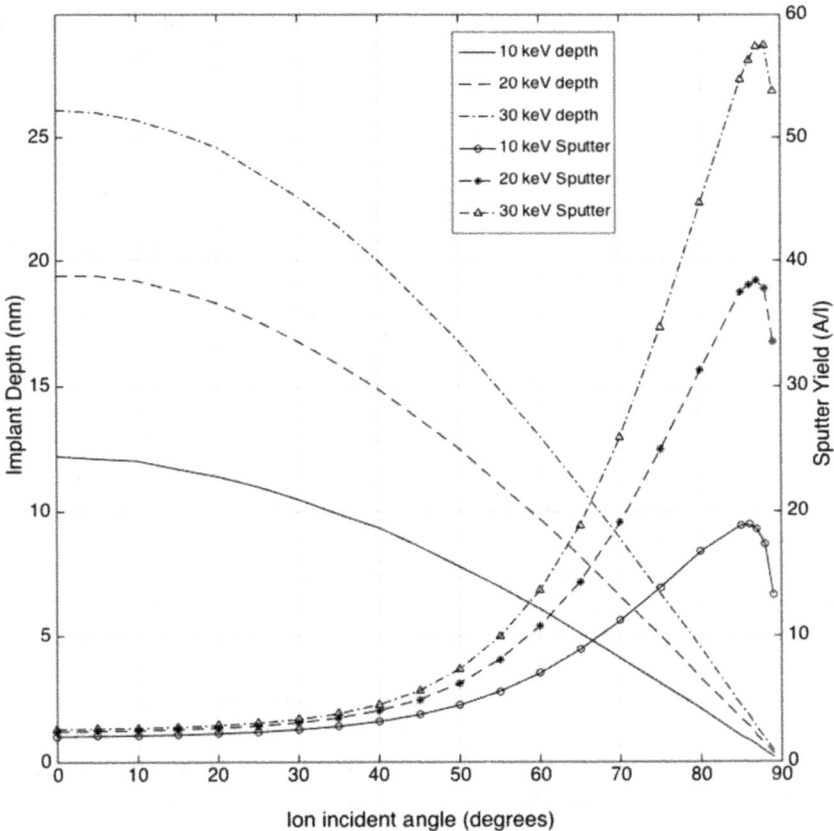

Figure 3.5. Ion ranges and sputter yields as a function of ion incident angle for three ion energies of Ga ions in an Si target.

different ion energies. At normal incidence, as previously described, there is only a small difference in sputter yield as a function of ion energy, with the range differing by a factor of about two. As the incident angle is increased the ion range diminishes rapidly and the sputter yield increases by more than an order of magnitude (at 30 keV) as the damage cascade becomes located much closer to the sample surface. Finally, the sputter yield begins to drop rapidly again as the incident angle becomes so shallow that Ga ions will begin to glance from the surface without transferring much of their energy to the sample. One further consequence of high incident angle is that even ions that enter the sample have a far higher probability of scattering back to the surface and the number of Ga ions retained in the material will fall significantly.

While the sputter yield of any given material is dependent upon the species, incident energy and incident angle of the impinging ions, this is an incomplete picture. The volume of material that can be removed in a given time by an ion beam is also dependent upon the flux of ions, i.e. the beam current. For gas field-ionisation instruments, beam currents have a working range from 0.1 to 100 pA and given that these instruments use the lightest species, sputtered volumes are correspondingly very small. This does not detract from their usefulness in niche nanofabrication applications, where extremely low volume removal is desirable, for example [10]. However, for volume removal on the scale of μm or greater both higher mass species and more importantly significantly higher currents are required. With currents ranging from 0.1 pA to 100 nA, the Ga LMIS instruments can remove useful volumes of material in a matter of minutes. For example, a Ga ion beam can easily produce a TEM foil of 10 μm width and 5 μm depth from a bulk sample in under an hour, a process involving several steps, including physical removal of the sample and its attachment to a TEM grid. If a region of interest in a sample becomes much larger than a few tens of μm then even Ga instruments become unviable to remove sufficient material in a reasonable time. Fortunately, the plasma-based instruments can operate with beam currents of up to several μA and, combined with their use of high mass noble gas ion species, can easily sputter volumes 100 times larger than a Ga-based instrument in an equivalent time with only a modest loss of resolution.

In much the same way that ion imaging produces grain contrast in polycrystalline samples, so sputter yield also varies with local crystallinity. The ion range will vary with orientation of individual grains and result in preferential milling (i.e. higher sputter yield) of grains that have the shortest ion range. An example of this is shown in figure 3.6, where a portion of an Nb thin film is being sputtered with a Ga ion beam. As can be seen, as milling progresses some grains have completely sputtered away, exposing the underlying substrate, whereas most of the film, while much thinner, is still continuous. The prediction of range and sputter yield in polycrystalline materials is notoriously difficult and calculations of these values for materials tend to be based on the assumption that the sample is amorphous. Furthermore, some materials will amorphise in the damage cascade much more readily than others, resulting in a change in sputter yield in the early stages of ion exposure. In polycrystalline materials the calculations of sputter yield and range should be considered as average values for all orientations.

In addition to the visible sample effects, such as sputtering, considerable sub-surface activity is also ongoing during ion irradiation. The exact nature of this

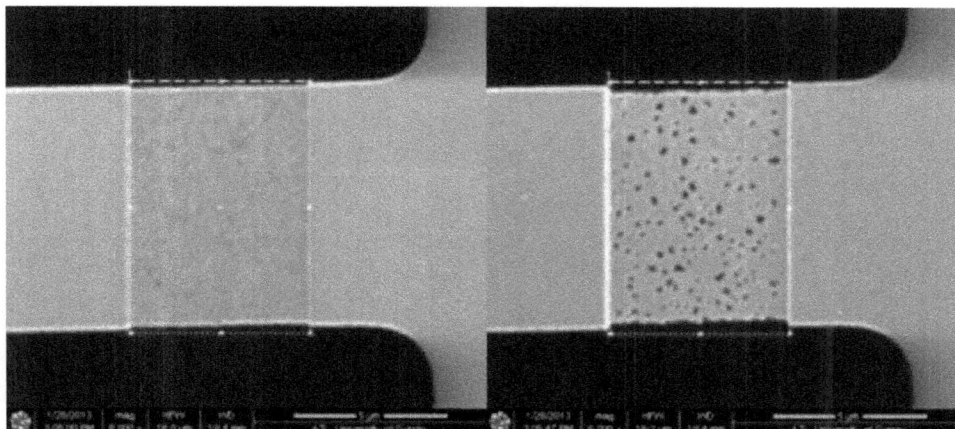

Figure 3.6. Preferential ion milling of a Nb thin film. Ion images of a thin Nb film during Ga ion milling. A short time after ion milling commences some contrast is visible in the grains of the film (left). As milling progresses some grains are sputtered preferentially, revealing the underlying substrate (right).

activity depends on the material, the ion species and the beam conditions, along with grain orientation in the case of crystalline materials, but generation of defects in the sample, such as vacancies, substitution of target atoms with the ion species in lattice sites, mixing of species across interfaces and even alloying is possible. A single ion, of the heavier species, can easily generate several hundred vacancies per ion during ion milling. As an example, figure 3.7 shows a calculation of predicted lattice damage as a function of ion Ga dose. For the reader who may be unfamiliar with these values, doses of the order of a few million ions per square micrometer can be considered very low and at these levels almost no sputtering effects would be visible in the instrument. Depending on the specimen material and range, it is quite possible for the ion beam to completely amorphise the sample to a depth up to a few 10s of nm at even quite modest doses; however, in the case of metals in particular, substantial vacancy healing would be expected to occur in the damage cascade and the sample will tolerate high doses while retaining crystallinity.

Turning to alloying of the sample and the ion species, this of course is not a problem with the noble gas species, but figure 3.8 shows the phase diagram for Sn/Ga and is a good example of where substantial problems occur if alloying is possible. At 100% purity Sn has a melting point of 505 K but the addition of only 8.4% (atomic) Ga reduces this melting point for the newly formed alloy to very close to room temperature. When one considers the ion range and likely dose of Ga required to sputter sufficient Sn, for example to study microstructure, achieving 8.4 atomic per cent in the near surface region of the sample is quite realistic. While sample heating by the beam is limited to the damage cascade [3] and is expected to be negligible in terms of the bulk, following Ga ion milling of pure Sn at room temperature the sample does give the appearance of having undergone surface melting. Other elements known to cause problems due to Ga ion implantation are predominantly the other post-transition metals, such as Al, In and III–V compound

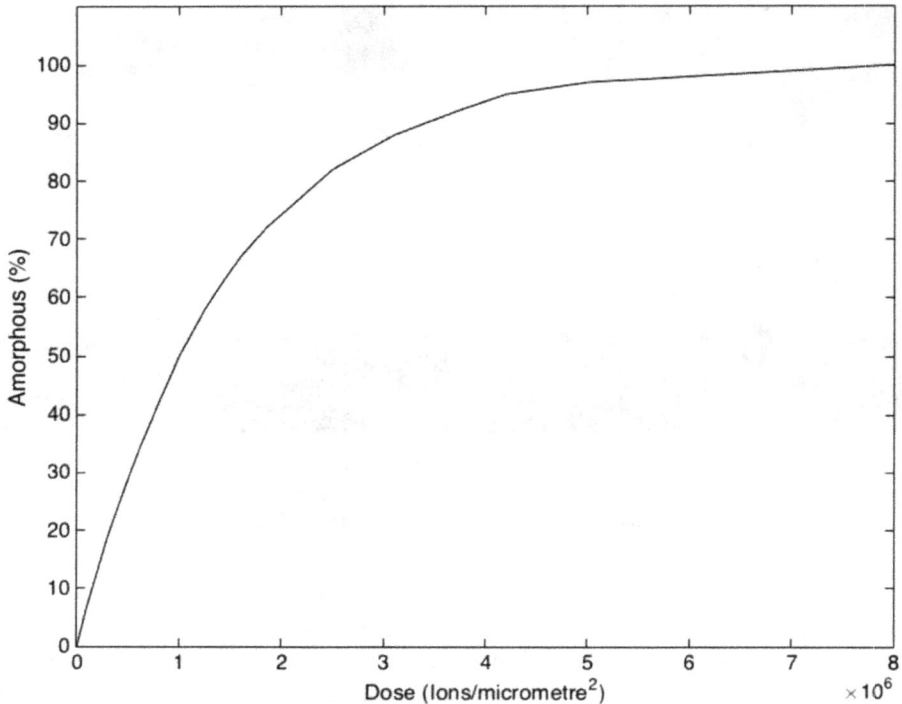

Figure 3.7. Progression of lattice damage in Si as a function of increasing 30 keV Ga, normal incidence, ion dose.

semiconductors such as indium antimonide, all cases where several low melting point alloys can form with the modest addition of Ga.

During the process of sputtering the ejected material will to a large extent be removed from the milling region by a combination of the instrument vacuum system and redeposition of the material on the instrument chamber walls and other internal components. Even allowing for the lifetime of a typical instrument, the sputtered volumes are very small and present no problem in this respect (except for possible small amounts of sputtered toxic materials that may be present when servicing pumps, etc, where care should be taken). However, in certain circumstances significant amounts of sputtered material can redeposit back on the sample in close proximity to the ion-milled region. This material will be amorphous in nature and may redeposit over ion-milled faces prepared for TEM examination, or may lead to electrical shorts in milled devices if it is conductive. Polymeric materials in particular can be prone to redeposition problems and the use of gas-assisted milling (chapter 2, section 2.4) can alleviate the problem. Generally, redeposition is more visible when milling cross-sections as the ejected material has a tendency to deposit on the opposite face to the ion-milled feature being produced. Figure 3.9 shows a large ion-milled volume being removed around a corrosion pit in steel to study its volume and local chemistry, and the extent of cracking that has originated from the pit. As milling has progressed, a substantial volume of sputtered material has redeposited on the adjacent face to the exposed corrosion pit.

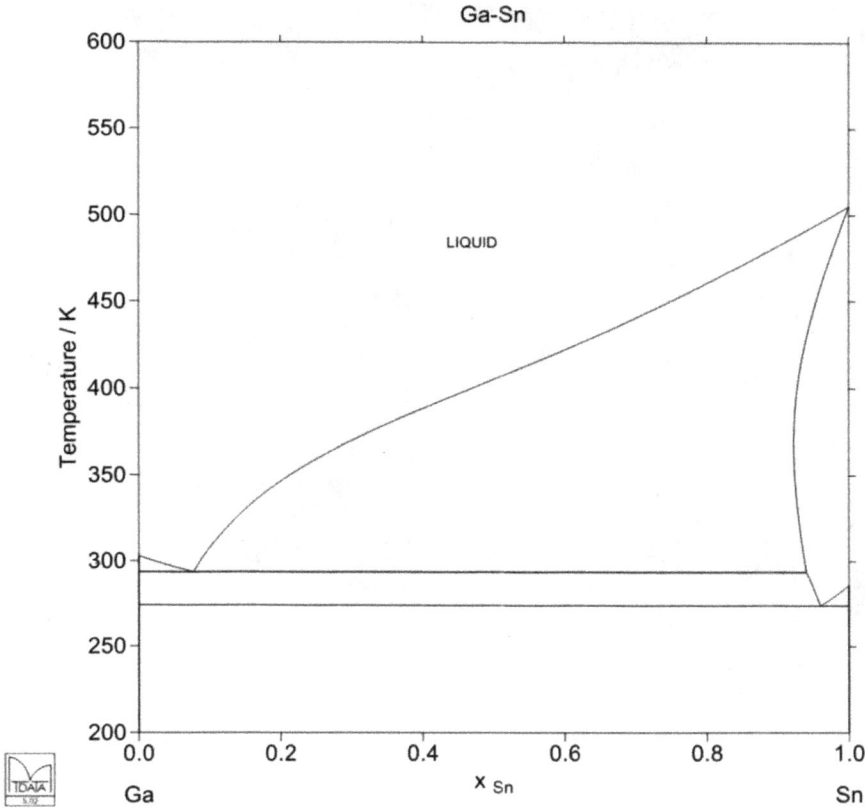

Figure 3.8. Phase diagram for Sn/Ga showing low melting eutectic point at 8.4% Ga.

3.4 Software to approximate ion range, damage and sputter yield

All the plots of range and sputter yield given in this chapter are derived from theoretical models. Establishing these parameters experimentally is very difficult to carry out with any certainty considering the vast range of materials, and combinations of materials possible in a single sample. It is simply not practical to establish these values experimentally for even a fraction of all possible combinations. As a result of this, a large body of work has been carried out that studies the theory of ion–solid interactions and several tools have become available that are of great use to the FIB operator. Essentially, these models aim to determine the energy deposition that takes place when an energetic ion impacts on a specimen and is either recoiled or stopped by the specimen. The deposition of energy at each scattering event can result in transfer of energy to sample atoms, which may be ejected from the sample surface (sputtered), or may in turn transfer their energy to other sample atoms, resulting in the generation of vacancies and secondary electrons. Once all of the available energy of the ion is given up in a combination of electronic and (far more dominant) nuclear energy loss events, then the ion will come to rest and its range will be determined. The statistical

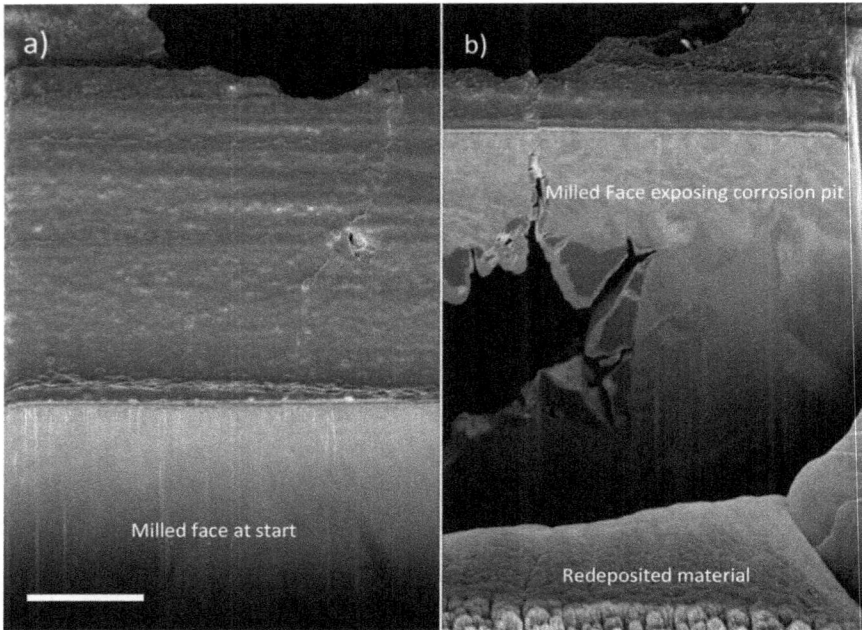

Figure 3.9. Split image of an ion-milled cross-section of a corrosion pit in stainless steel. The left side of the image (*a*) shows half of the sample in the early stages of sectioning and the right image (*b*) (taken some considerable time later) shows the other half. The extent of the pit has been revealed by the removal of several μm of material and shows considerable redeposition on the opposite face to the milled section. The scale bar indicates 2 μm. Reproduced with permission from Helen Jones (NPL).

nature of each scattering event results in a symmetrical Gaussian distribution of stopping ranges around the mean, termed the range straggle. During scattering events the ion can, of course, move in three dimensions and its lateral deviation from the point of impact again follows a Gaussian distribution, termed the lateral straggle.

Two computer programs are currently freely available to model the stopping of ions and subsequent energy transfer in solids and associated sputtering. The Stopping and Range of Ions in Matter (SRIM) [4] utilises Monte Carlo simulation and the Surrey University Sputter Profile from Energy Deposition programme (SUPRE) [5] uses a numerical approach. SRIM calculates the three-dimensional resting location and associated damage cascade ion by ion in simple pure materials, alloys, compounds and complex multilayer materials, and can calculate the ion location, dislocation density, interfacial mixing of species and sputter yield of individual components in multicomponent systems. An assumption is made in all cases that the material is amorphous, but density adjustments can be made if it is known that the system to be modelled is not fully dense. By assuming the material is amorphous no account of ion channeling due to material crystallinity can be made, but SRIM can be regarded as giving averaged values for all orientations and it has been found to be fairly robust for the FIB process in both amorphous and crystalline materials. It should be remembered that by calculating ion by ion a significant

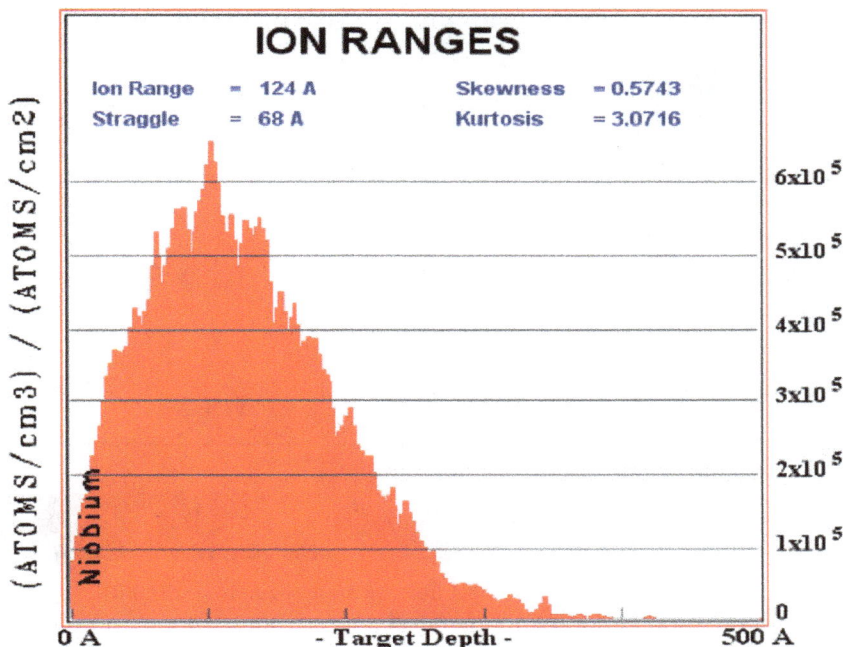

Figure 3.10. Typical two-dimensional plot of an ion range (30 kV Ga in Nb) produced by SRIM.

number of ions need to be modelled to obtain significant distribution statistics and accurate sputter yields. SRIM can offer differing levels of calculation, from simple distributions of ions to full damage cascade with surface sputtering. Calculation times vary significantly for these different levels of complexity, but even large numbers of ions can be modelled on modern computers in minutes. The output from SRIM can be in the form of graphic display in real time, simple text files, or a multitude of different plots of many different aspects of the ion implant process in both two dimensions for each Cartesian direction and in three dimensions. An example of a simple two-dimensional plot showing an ion range of 30 kV Ga ions in an Nb target is shown in figure 3.10. For a full and very detailed description of the underpinning mathematics and methodology the reader is directed to [11].

By comparison to SRIM, SUSPRE uses a numerical solution to the Boltzmann transport equation and is based on the projected range algorithm (PRAL) [12]. It is designed to calculate an approximate solution for the implantation range profiles of any ion in any target material, additionally calculating both the energy deposition [13] and from this expected sputter yield [14]. With a very simple to use graphical interface, it produces two-dimensional plots of ion range, energy deposition and damage, along with values for mean range, straggle and sputter yield. Figure 3.11 shows a typical SUPSRE screen output. As with SRIM, SUPSRE assumes the target is amorphous, but includes options to modify sample density and can accommodate compounds and alloys. Unlike SRIM, it cannot calculate ion range and damage in multilayer materials.

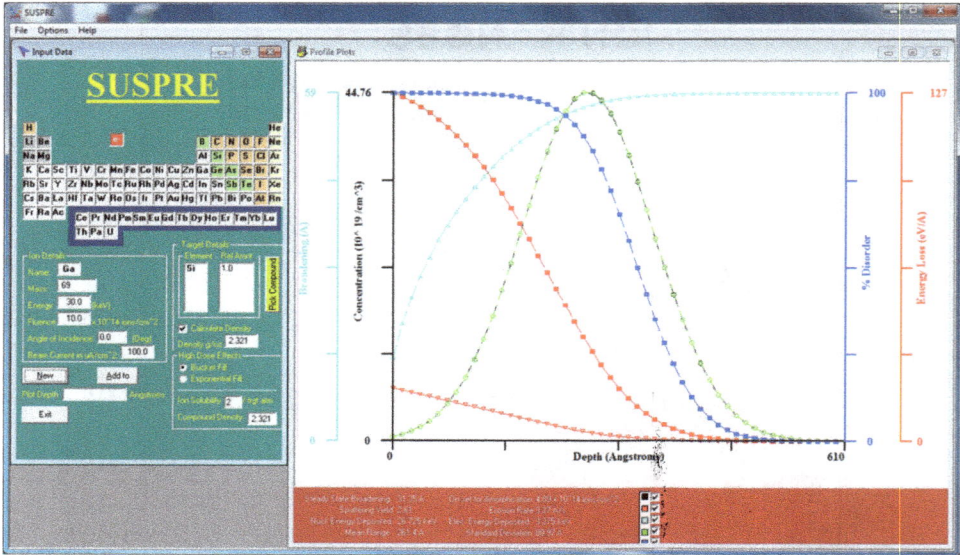

Figure 3.11. Typical screen output from the modelling software SUSPRE.

In addition to SRIM and SUSPRE, other computer codes are also available that simulate range and sputtering in crystalline materials. Two examples of these are the MARLOWE code [15] and ACOCT [16]. However, the complexity of the calculations, restrictions on the complexity of the system that can be modelled and relative difficulty of use when compared to the extreme simplicity of SRIM and SUSPRE limits their usefulness for everyday operators of FIB systems.

References

[1] Nastasi M, Mayer J and Hirvonen J K 2004 *Ion–Solid Interactions: Fundamentals and Applications (Cambridge Solid State Science Series)* (Cambridge: Cambridge University Press)

[2] Yao N (ed) 2007 *Focussed Ion Beam Systems* (Cambridge: Cambridge University Press)

[3] Giannuzzi L A and Stevie F A (ed) 2005 *Introduction to Focussed Ion Beams* (Berlin: Springer)

[4] www.srim.org

[5] www.surrey.ac.uk/ati/ibc/research/modelling_simulation/suspre.htm

[6] Seiler H J 1983 Secondary electron emission in the scanning electron microscope *J. Appl. Phys.* **54** R1

[7] Cazaux J 1999 Some considerations on the secondary electron emission, delta, from e(-) irradiated insulators *J. Appl. Phys.* **85** 1137–47

[8] Orloff J, Utlaut M and Swanson L 2003 *High Resolution Focused Ion Beams: FIB and its Applications* (Dordrecht: Kluwer Academic)

[9] Ishitani T, Madokoro Y, Nakagawa M and Ohya K 2002 Origins of material contrast in scanning ion microscope images *J. Electron Microsc. (Tokyo)* **51** 207–13

[10] Bell D C, Lemme M, Stern L A, Williams J R and Marcus C M 2009 Precision cutting and patterning of graphene with helium ions *J. Vac. Sci. Technol.* B **6** 2755–8

[11] Ziegler J F and Biersack J P 2008 *SRIM–The Stopping and Range of Ions in Matter* (Raleigh, NC: Lulu Press)

[12] Biersack J P 1981 Calculation of projected ranges—analytical solutions and a simple general algorithm *Nucl. Instrum. Meths* **182/183** 199–206

[13] Gibbons J F 1972 Ion implantation in semiconductors—part II: Damage production and annealing *Proc. IEEE* **60** 1062–96

[14] Sigmund P 1969 Theory of sputtering. I. Sputtering yield of amorphous and polycrystalline targets *Phys. Rev.* **184** 383–415

[15] Robinson M T 1989 Slowing-down time of energetic atoms in solids *Phys. Rev.* B **40** 10717–26

[16] Yamamura Y and Takeuchi W 1987 Monocrystal sputtering by the computer simulation code ACOCT *Nucl. Instrum. Meths* B **29** 461–70

Chapter 4

Focused ion beam—materials science applications

4.1 Overview

By far the greatest use of FIB systems is in the characterisation of materials and their properties. From simple cross-sections to study defects in electronic devices, preparing TEM lamella for studying the arrangements of atoms in crystals, or determining properties such as stress in films, FIB has found widespread and still significantly growing use around the globe in hundreds of university, government and industrial laboratories. The ability to slice open a sample at very precise locations and see the internal structure, on a scale ranging from atomic resolution (in samples prepared for TEM) to hundreds of μm on such a huge range of materials is unsurpassed by any other technique currently known. This precision, combined with the fact that the sample can then be examined by other techniques, such as TEM or EBSD, provides a wealth of opportunities for the study of the metrology of materials. This chapter discusses key areas where FIB is used in the study of materials. It is not exhaustive and the field is still very dynamic, with new methods and techniques being added frequently. However, the topics covered represent the most commonly found areas where FIB is used in materials studies.

4.2 TEM foils and cross-sectioning

TEM is well established as one of only a few techniques that can image samples on the atomic scale non-destructively. Whether it is imaging atomic structure, dislocations, interfaces, or local chemistry, TEM is a very common and widespread technique. One of the limitations of TEM, however, is the preparation of samples. As a general rule, specimens must be smaller than 3 mm diameter and thin enough that the beam can penetrate the sample. In practice, this can mean that samples can be as thin as a few nm, and at most typically 1 μm, depending on sample type and density. Many techniques have been developed and sometimes combined to produce

doi:10.1088/978-1-6817-4084-3ch4

Figure 4.1. TEM lamella from a lanthanum hexaboride crystal prepared on a FEI dual beam FIB system, showing the atomic resolution and structure of the crystal, visible in a FEI Titan G2 TEM (inset). The location of the lanthanum atoms in the crystal lattice is depicted in red and the boron in blue. Reproduced with permission from the FEI Company.

these samples, including simple mechanical polishing, high incident angle ion polishing, chemical polishing, microtomes and exfoliation, amongst others [1].

The FIB lends itself incredibly well to the preparation of TEM samples [2]. The ability to remove material from a sample at very precise locations makes accessing specific regions possible with an ease that is often unobtainable using other methods. Essentially, a FIB lamella in bulk samples comprises two cross-sections cut back to back with the remaining material in-between the material to be studied. This lamella will need to be lifted from the sample and attached to a suitable TEM grid and tools for both *ex situ* or more commonly *in situ* lift-out are required, and are discussed in section 2.5. Once removed, the lamella can be further thinned while attached to the grid using both lower ion beam accelerating voltage and high incident angle. This combination gives rise to high removal rates and a higher probability of the incident ions scattering from the sample, improving the quality of the sample by removing damage and the implanted ions from the initial stages of production. With this method and a suitable sample it is possible to achieve a sub 10 nm thinness, with sub 100 nm being achieved easily in many materials. Shown in figure 4.1 is an example of a thin TEM lamella cut from a lanthanum hexaboride crystal and the atomic resolution that can be attained in the TEM from FIB-prepared samples. For a discussion of the many techniques available in TEM, the interested reader is directed to [1]. Depending on the ion species, and equally importantly the available beam current, the maximum area that can be studied by a FIB-prepared lamella ranges from a few μm^2 upwards for a Ga FIB to hundreds of μm^2 for a Xe plasma FIB. Owing to the negligible sputter yield of He and the very low yield of Ne, the gas field-ion sources are not practical for lamella production, but in principle Ne could be used for final polishing.

As has been discussed, a TEM lamella is produced in bulk samples by removing two volumes of material either side of the region of interest. A simple variation of this approach is to sputter a single volume and image the resulting cross-section with either the ion column, or in the case of dual beam instruments the SEM column. Such a simple technique could, for example, be a very rapid method for determining film thickness, as shown in figure 1.2, but again the targeted nature of the FIB can create a cross-section at very precise locations that can be imaged by the many available detectors, such as secondary electrons or ions, backscattered electrons, EDX and EBSD. The latter two techniques, however, may require substantial clearance of surrounding material as the geometry of a simple cross-section may mask the detector from the signal, particularly in the case of EBSD. As with all imaging methods, the absolute resolution is determined by particular beam conditions, such as beam spot size, source brightness and the interaction volume of the beam with the sample. As a general rule, it could be considered that SEM imaging with secondary electrons would give close to single nm resolution, secondary ions a few nm, EDX sub-μm and EBSD a few tens of nm; however, the large number of variables of both instrument conditions and sample can give rise to large variation in these values.

4.3 Three-dimensional reconstruction

Extending on beyond simple cross-sectional imaging, FIB systems can also explore the three-dimensional nature of samples—FIB tomography. Multiple slices of a material can be milled sequentially through a region or feature of interest to acquire sub-surface and cross-sectional information (figure 4.2). In dual beam systems where the second column is usually an electron beam, SEM images of the cross-section can be captured between milled slices and reconstructed in three dimensions into a volume to visualise the milled-away region. Such an approach can give valuable information about microscale structures that is unobtainable by other methods [3–5] and can be applied to simple imaging, EDX or EBSD data from each slice.

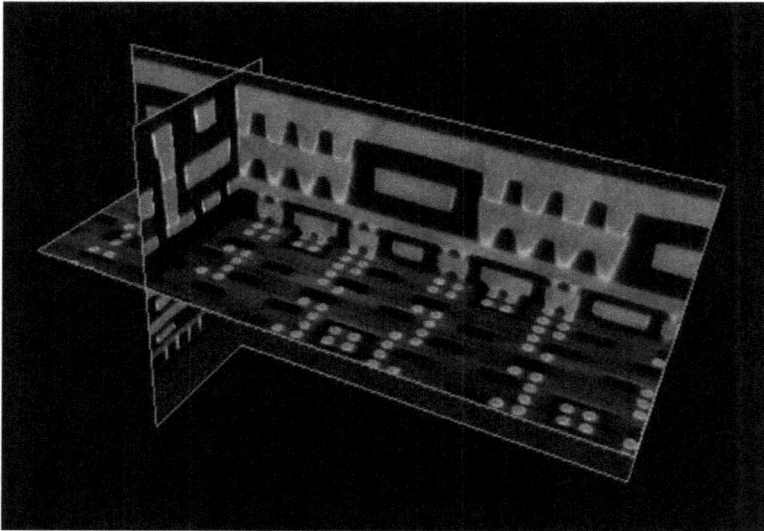

Figure 4.2. Three-dimensional reconstruction from FIB slices of an electronic device interconnect structure captured on a dual beam system using backscatter electrons.

Individual slice thickness, while varying with ion species, beam current and sample, can be as thin as 20 nm (in principle less), and can be thought of as the minimum resolution attainable. However, even if the sampling volume of the particular imaging technique used can achieve this (for example, with EDX it would not be possible as the sampling volume can be on the scale of hundreds of nm or more), when reconstructing three-dimensional microstructures there are corrections that need to be considered to reconstruct the volume accurately. The individual slice images in the X–Y plane must be aligned accurately with each other, while in the Z direction corrections for the actual slice thickness must be made. Slice alignment in the X and Y planes can be made manually, or using automated software routines [6, 7] with a wide range of algorithms that can be limited to translation, or extended to allow shearing or stretching to compensate for drift during acquisition. The required slice thickness is set as one of the parameters of the milling, and this set value is commonly taken as the depth to be used with the X–Y pixel size to create cubic voxels in the three-dimensional reconstruction [8–11].

However, the use of the set slice thickness values and algorithms to correct for errors ignores the possibility of unknown errors in the actual slice thickness achieved [11]. Inaccuracies can arise from errors in calibration, giving a constant error in all slices, or from variations between slices (which often add to misalignment in X and Y) caused by, for example, charging effects [12, 13], beam heating and the addition of stage movements between imaging and milling operations, as is often required with EBSD in particular. Measurement of the actual slice thickness for every slice, using comparisons between the location of the new leading edge of a slice and the previous slice, can be difficult. Examples of the inaccuracies resulting from no, or limited, measurement of the actual slice thickness are discussed by Mingard *et al* [11].

Figure 4.3 shows simple crossed line fiducial marks milled at a known angle to the milling plane, on the top surface of a sample. Marker schemes such as this have been used in several studies for alignment [14] or assessment of average slice thickness, with there even being a patent filed for individual slice thickness based on this geometry [15]. All these methods will, of course, only give the thickness at the surface based on the slice leading-edges, providing no evidence of how parallel the slices might be. However, the addition of fiducial marks on the top surface to measure slice thickness and drift during milling is a *minimum* requirement to assess likely errors that may have occurred during the milling operation. Without these marker systems, any three-dimensional reconstruction is likely to be in error, with both the magnitude and direction of any error indeterminable, except in very special cases, for example, where sectioning a perfectly spherical object, or a structure where the geometry is already known and the user is looking for defects (see figure 4.2).

Determining the errors that may be occurring during slicing is extremely difficult. Not only can slice thicknesses vary from one to the next, but the slices can be twisted, tilted or warped with respect to each other. A full study of these errors has been demonstrated in [16], where a variant of the previously described marker system was extended into the third dimension. This sample is a classical example of a metrological artefact where the precise geometry of the specimen is known in advance and can be used to subsequently determine error that would otherwise be

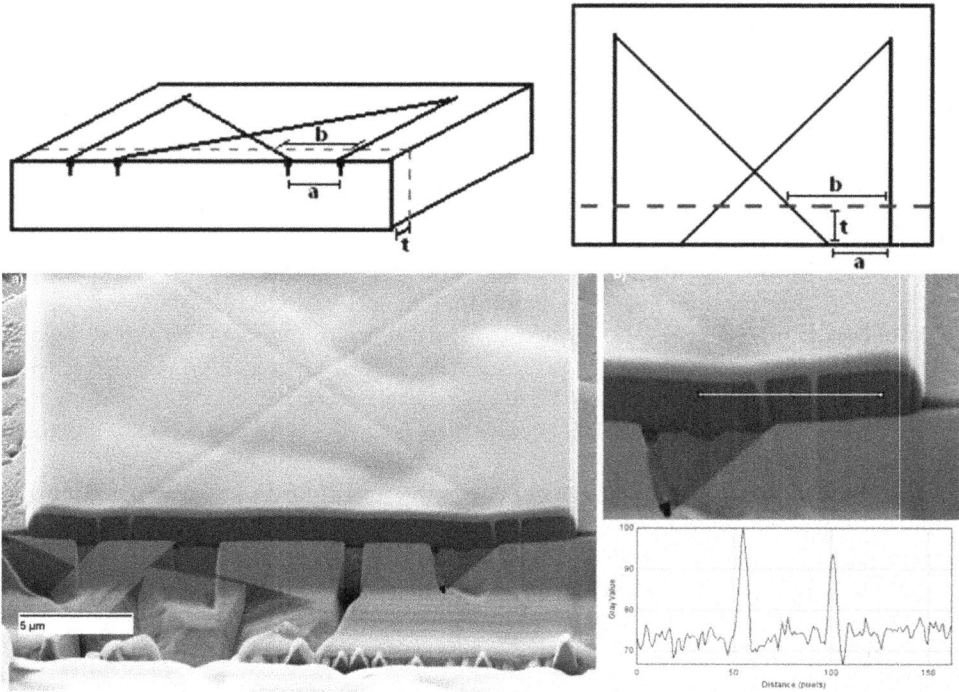

Figure 4.3. Schematic diagram (top) of a fiducial marker system that could be used to asses slice thickness, with a cross and two parallel lines ion milled into the surface of the sample. As slices are removed the fiducial is milled back by thickness 't'. If the line geometry is known, the thickness can be assessed by careful measurement of the change in length of the lines and and the distance from the parallel ones to the intercept of the cross (b). The SEM micrographs (bottom) show the marker system in practice. In this example, carbon (from the FIB GIS) has been used to put a low contrast pad on the surface of the sample. This in turn has been ion milled with the cross and parallel line and subsequently coated with Pt (GIS) that gives a high contrast compared to the C. Once sliced, the marker lines are clearly visible and can be measured with high accuracy (inset), determining the slice thickness that has been removed. Such a system gives a simple measure of slice thickness and the possible non-parallelism of the slices. Reproduced with permission from [16].

unobtainable from an unknown geometry. Figures 4.4 and 4.5 show the methodology and construction of this artefact, all carried out in an FIB instrument (with a modified electron beam lithography system to enable precise alignment of the layers) using standard GIS chemistries and simple FIB patterning. These samples are subsequently sliced with FIB under different conditions with both FIB and SEM images captured at the end of each slice. The marker system can then be measured from the SEM images as the slices progress, giving a three-dimensional measure of a 10 μm cube containing up to 48 measurement points for each slice over the whole face (each slice has eight points along each layer interface, but only six will be visible at at any time, and eight patterned layers).

Figure 4.6 shows a data set obtained from one of these stack artefacts. The three plots show the average slice thickness in pixels (a) or nm ((b) and (c)) at various points in the stack. Also shown in figure 4.6(a) is the target thickness (thick line) and the running

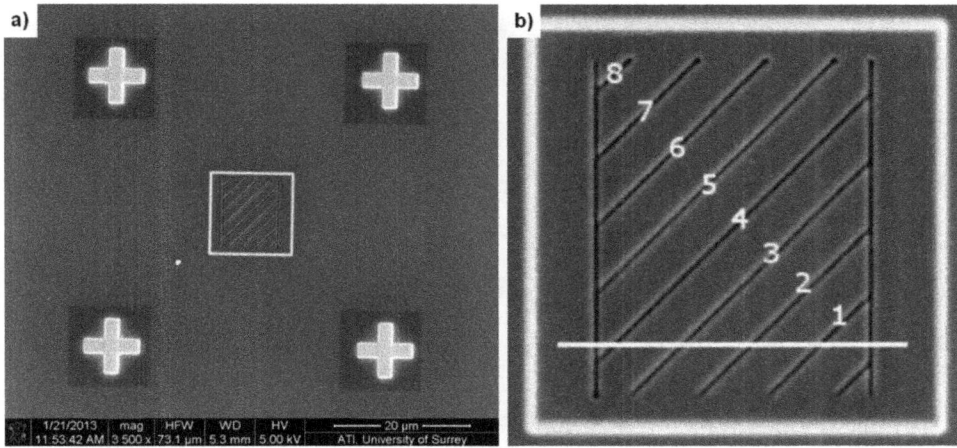

Figure 4.4. SEM micrographs of the initial stages of the slice artefact construction. (*a*) Low magnification image showing the alignment crosses used to carefully align each layer and the first pattern. (*b*) Higher magnification image of the parallel line pattern. Reproduced with permission from [16].

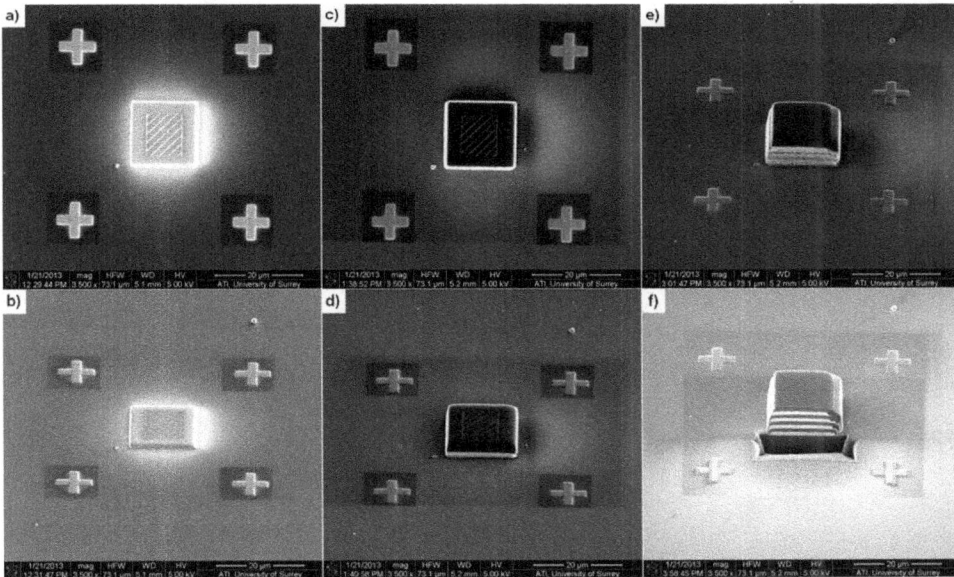

Figure 4.5. Series of images showing the subsequent construction of the slice artefact. Alternating layers of Pt and C are deposited with the FIB GIS (*a*)–(*e*) with the parallel line patterned on each layer, and a cross-section of the stack of layers showing the alternating high and low contrast layers. Reproduced with permission from [16].

average of thickness as milling progressed. As can be seen, significant errors exist and the first few slices were only half as thick as the target, while the next few slices were noticeably thicker. The thicknesses then oscillated around the target, with very few actually at the correct thickness. Also, looking at the data from different layers in the

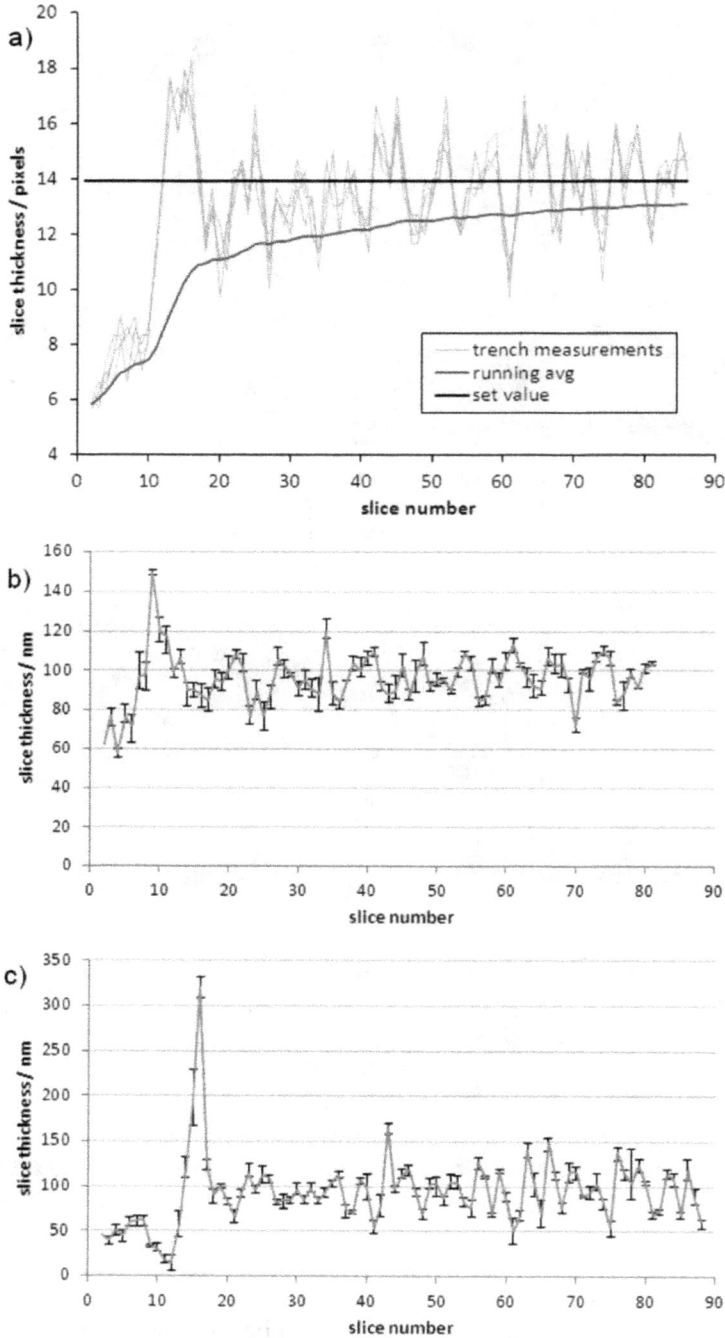

Figure 4.6. (*a*) Raw slice thickness data for the seventh layer of all markers with set slice thickness value and running average. Slice thicknesses for the (*b*) top and (*c*) bottom layer of the stack with error bars showing uncertainty (range). Note that the scale on the *y* axis has intervals that are half the values for (*b*) the top layer compared to those in (*c*) the bottom layer. Reproduced with permission from [16].

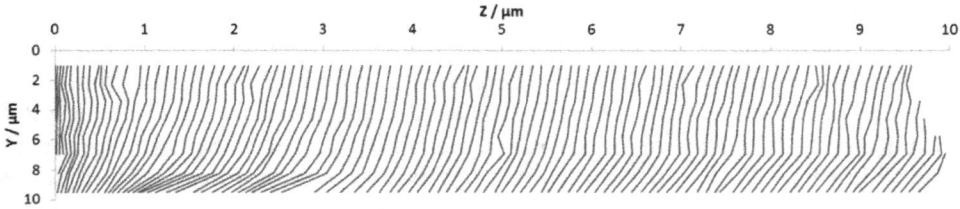

Figure 4.7. Extrapolated slice thickness to represent the vertical section through the centre of the stack, showing the slice shape as viewed perpendicularly to the milling direction. Reproduced with permission from [16].

stack it is clear that the ion-milled faces are not parallel and have some distortion. Using this method with several samples, sliced under differing conditions and slice thickness, on two different FIBs from different manufacturers, the users found no slice and view run yielded reproducible slices and significant, but apparently unsystematic, errors were present in every three-dimensional reconstruction. Furthermore, it was found on odd occasions that 'phantom slices' occurred, i.e. the instrument ran as normal but a slice was misallocated to such an extent that no milling occurred on the edge of the sample. In these cases there was also some evidence to suggest that redeposited material from milling the slice in the incorrect location actually degraded the SEM image quality by covering detail on the sample face.

The advantage of having multiple markers throughout the artefact also enables a study of the full geometry of individual slices rather than just establishing the thickness of each slice with respect to the surface. While this would normally not be possible on an unknown sample, it further shows the significant errors that can occur. Figure 4.7 shows a reconstruction of the shape of the ion-milled face by taking the values for the central markers for each slice. Essentially, this is a 'side-on' view of the slices and shows how most of the slices have some tilt, with several having noticeable curvature over the first third of the set, which diminishes, but increases again towards the end.

While it has been shown that significant errors can exist in the three-dimensional reconstruction of samples, this should not detract from the usefulness of the method and the fact that the site-specific nature of FIB sectioning provides a route to study the internal structure of unknown objects and structures that may not be possible using any other method. As long as the FIB user is prepared to accept (and equally importantly, state) the errors of their measurement, and if at all possible, use a surface marker system to minimise these errors, three-dimensional cross-sectioning is an immensely powerful technique. Furthermore, some software used in three-dimensional reconstruction can also align stacks of images based on edge or particle detection methods, and is particularly useful in three-dimensional EBSD, where grain boundaries provide natural edges for software to detect. However, some care is needed to understand the methods the software uses and it should not be treated as a panacea that will eliminate errors introduced by both FIB slicing and SEM imaging, as they are often complex and connected in an unknown way.

4.4 Mechanical testing

While FIB is used extensively as an imaging tool or as a means to prepare samples for imaging and studying chemistry and structure, it is also finding widespread use as a means to manufacture specimens for miniaturised mechanical testing. In cases where only very small amounts of sample exist (for example, alloy development), or the samples are inherently small, or where there is a desire to study deformation at the micro- and nanoscale (for example, plasticity on the scale of individual crystals in a bulk sample), FIB is finding use as a machining tool to make such specimens.

Until fairly recently, some uncertainty has existed over the effects of sample size on the strength and plasticity of mechanical tests specimens. This uncertainty was in most cases hypothetical, with samples on the scale of micrometres, as the vast majority of mechanical tests specimens are on the scale of at least mm, and were except in very special cases (such as specially grown micro scale whiskers) impossible to produce and test at such reduced dimensions. FIB has, however, now enabled specimens to be produced at this massively reduced scale both quickly and in a targeted way. One of the simplest geometries that a FIB can produce for mechanical testing is a simple uniaxial compression pillar (an example is shown in figure 4.8). Such samples are subsequently tested with flat punches on either *in situ* or *ex situ* nano-indent test machines. Samples of this geometry were used to test the effects of sample size on strength and plasticity [17]. Here it was found that the overall sample dimensions artificially limit the length scales available for plastic processes and show significant size effects with even relatively large (by FIB standards) sample dimensions. Rather than a smooth transition to plastic flow samples (of Ni and Ni superalloy), they exhibited a burst of activity, with large jumps in strain that on occasion approached 20%. This behaviour was attributed to the samples being very small single crystals with limited capacity to accommodate defects that

Figure 4.8. Uniaxial micro-compression pillar ion milled from tungsten carbide and crushed *in situ* in the FIB instrument and showing clear slips bands as the sample deforms under load. Reproduced with permission from Helen Jones (NPL).

rapidly multiply and subsequently saturate the test pieces. Unlike larger crystals or polycrystalline materials, the reduced scope to accommodate damage results in much higher yield stresses, little work hardening and behaviour that is more akin to a single crystal metal whiskers. The size effect on mechanical behaviour of this reduced size pillar can, of course, make translating data obtained this way to large scale bulk materials far from straightforward. However, similar tests have been been carried out on nanocrystalline thin films (used in electronic applications) whose dimensions are on a similar scale to the compression pillars and so the obtained values can be considered to present real samples [18].

Other geometries of test sample are also possible with FIB, for example micro tensile on pure copper [19] and larger (100 μm gauge length) polycrystalline tensile bars combined with EDX/EBSD [20], the latter of these showing that test samples of this size are perfectly feasible as an alternative to large-scale macroscopic samples. It should be said, however, that the samples in this study were largely prepared *ex situ* by large-area ion etching using a Si mask and were only finished (and measured) in the FIB as it was a Ga-based instrument. If Xe plasma FIB had been used it would have been perfectly feasible to carry out the whole process in the instrument and produce test pieces of similar dimension using FIB alone.

A very novel experiment and one that shows how FIB can be used in very unique ways on specimens that could not be produced in any other instrument is demonstrated by Barber *et al* [21]. Figure 4.9 shows the details of this experiment on the tensile strength of limpet teeth. Here the tooth of the sample has been removed using FIB to first cut it from its root, then GIS has been used to attach it to an atomic force microscope (AFM) cantilever tip, where it is subsequently ion milled into a classical dog-bone geometry used in tensile testing. An AFM with load/displacement sensors is then used to pull the sample to destruction while measuring elongation and load (figure 4.10). Even allowing for the previously mentioned size effects in metallic samples and the fact that similar effects may be present, this study shows how mechanical tests on very challenging samples become feasible with FIB systems.

As was shown in chapter 3, irrespective of the ion species, energy and incident angle, damage to the crystal structure and implantation of the ion species (along with possible alloying with the specimen) is significant in FIB-exposed surfaces.

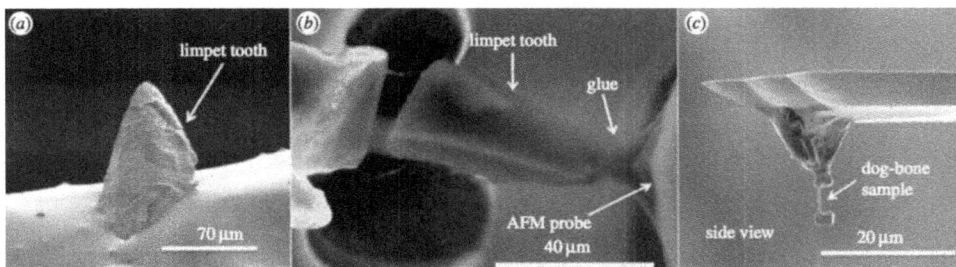

Figure 4.9. Procedure for constructing tensile specimen from limpet teeth. SEM images of (*a*) the limpet tooth, (*b*) attachment of a tooth fragment to an AFM cantilever using FIB and GIS and (*c*) the finished FIB-milled tensile test piece. (Reproduced from [21] under the Creative Commons (CC BY 4.0) licence).

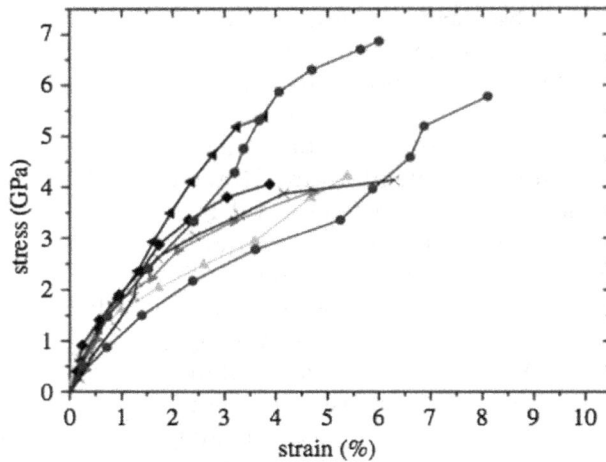

Figure 4.10. Plot of the stress–strain behaviour of individual limpet tooth samples, with a variety of lengths, tensile tested to failure using AFM. (Reproduced from [21] under the Creative Commons (CC BY 4.0) licence).

In this respect, the production of mechanical test pieces using FIB is no different and it should be expected that the effects of the ions may have an influence on a mechanical test piece that may be on the scale of only several μm. As Ga is the most commonly used source, studies have been carried looking at the ion implantation range and degree of damage of Ga on Cu [22, 23] and WC–Co [24], showing significant Ga concentrations near the surface along with substantial dislocation networks. Kienera *et al* [22] in particular have demonstrated the effects on reduction of both yield points and maximum yield stresses that ion damage introduces by annealing of FIB-milled specimens. The reduction in strengths in the unannealed samples is attributed to the presence of ion-induced defects facilitating further dislocation nucleation.

While it is without doubt that FIB offers a route to mechanical testing on a range of samples that were previously untestable, and further offers the chance to see in real time the effect of stress and yielding on the micro scale, it should be born in mind that mechanical data from these specimens may be affected by both size effects and ion damage when the samples are on this scale. The effects of reduced size increase apparent yield stresses, whereas ion damage does the opposite. These two effects do not fully cancel each other out and should be considered carefully when obtaining data by this method. If a Xe plasma FIB is available then a better strategy would be to fabricate samples that are of the order of 10 μm or more, where these effects diminish considerably.

4.5 Residual stress measurement and deformation

The measurement of residual stress in materials and components is of vital importance in many areas. Whenever a material is processed, for example, machined or joined, whenever a protective coating or film is applied, or even when a component is in service, the accumulation of residual stresses can exert significant influence on the properties and lifetime of the component. In many cases engineers go to great lengths to avoid the introduction of these stresses, and on other occasions deliberately engineer the stresses into the component. A simple example could be that of a thin film; these are often engineered to have compressive residual stresses, as this helps with adhesion of the film, whereas tensile stresses in such films are more likely to cause cracking and delamination. The measurement and determination of residual stress is therefore of significant importance to industry, to check components are manufactured correctly, or to determine their remaining service life, and several approaches have been developed over many years to establish residual stresses in materials [25].

Residual stress measurement falls into two categories: non-destructive and semi-destructive. Of the non-destructive methods, x-ray diffraction is by far the most widely used and best understood, but along with a related technique (neutron diffraction), it is largely limited to crystalline specimens, as these techniques look for distortions of the crystal lattice. Other non-destructive methods include determination of curvature, ultrasonic probing, EBSD and, where applicable, magnetic domain determination. The semi-destructive methods are not exclusive to crystalline materials and all rely on removing a small portion of material, often referred to as hole drilling, and actually measure the relaxation strain that occurs in the surrounding material, shown schematically in figure 4.11. The relaxation strain, once determined, can in turn be related to the pre-existing stress magnitude by numerical deformation modelling, as long as the Young's modulus of the material is known.

Typically, when carrying out hole drilling for stress determination the hole is macroscopic in size, often several mm in diameter and depth. The parameters available in ion beam milling, i.e. the energy and mass of the ion species, and the available beam currents, dictate that volumes of material, albeit at very precise locations,

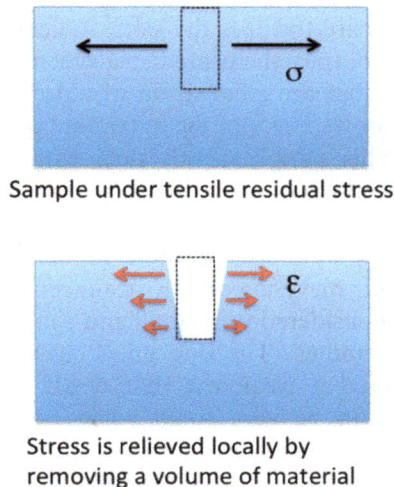

Sample under tensile residual stress

Stress is relieved locally by
removing a volume of material

Figure 4.11. Schematic diagram of semi-destructive residual stress determination. A cross-section of a sample with tensile residual stress, with a volume of material to be removed, denoted by the dashed box (top). On removing the section of material the sample undergoes relief strain in the surrounding area.

are microscopic. In recent years several studies have explored if hole drilling for stress determination can be pushed down into the microscopic regime using FIB [26–29]. What these studies have found is that not only is the technique viable, and in good agreement with other techniques, it also offers the ability to test samples that would be extremely difficult to measure with macroscopic techniques, for example, very thin films.

One minor limitation with stress determination in the FIB is that the holes are by definition very small and so too are the displacements that occur due to the relief strain. The only method for obtaining the displacement data is to use digital image correlation (DIC) software to compare images of the sample before and after the hole(s) is milled. Fortunately, DIC is capable of extracting sub-pixel level displacements in the images, where typically in an image an individual pixel may represent just a few nm. A further minor weakness is that as DIC is employed, there must be sufficient detail on the sample surface for the software to compare images. With samples where there is insufficient surface detail it is necessary to decorate the sample surface with a high density of particles or a pattern on a scale approximating to the pixel size. A simple method were a GIS is present is to use a randomised dot pattern and the beam to despot the pattern. On two column instruments use of the lectern is preferred as the ion beam could damage the sample and relieve some of the stress in the damage cascade.

An example of hole milling and DIC is given in figure 4.12, where two SEM images of a superalloy sample with an unknown residual stress are shown. The top image shows the sample before milling but after decoration with Pt GIS using the electron beam, with the centre image showing the sample after two slots have been milled with the FIB. The two slots have been milled repeatedly in increments of approximately 350 nm, with electron beam images captured after each milling step. The final panel of figure 4.12 shows a typical screen output from the DIC software, displaying the magnitude of displacement fields around the slots and showing that the displacement directions are towards the slots, indicating that it is compressive. As is evident from the DIC output, the peak displacements are less than two pixels in magnitude, demonstrating that the relief strain would not be determinable without the software. Figure 4.13 shows a plot of the maximum determined relief strain after each milling increment. This sample clearly demonstrates that at approximately 1.5 μm deep the relief strain has saturated and milling deeper sees no further increase. Two points to take from this observation are that firstly as a general rule the maximum depths of any slots or other hole geometry needs to be of a similar magnitude to the width to achieve maximum strain relief, and secondly with a Ga FIB these volumes can be considered very small and can be milled with even modest beam currents in a few minutes. The one great strength of the technique can, however, also be considered a weakness that needs some consideration if the sampling volume is likely to be very small when compared to the size of the specimen. Arguably, this is also the case with macroscopic hole drilling, but some caution should be applied to unknown specimens if they are very large compared to the hole size, as stress may not be uniform in the sample.

The limitation with milling simple slots is that they can only measure stress in a single direction. While it is possible to mill multiple slots at alternate orientations

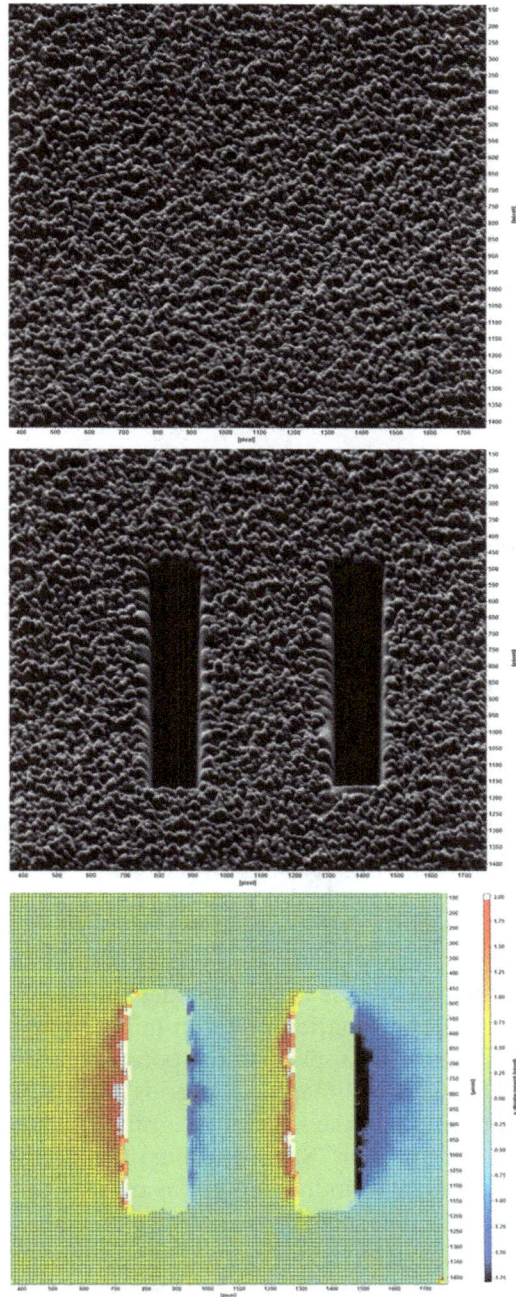

Figure 4.12. Example of two-slot stress determination in the FIB. The top panel shows an SEM image of the sample surface decorated with electron beam deposited platinum to give texture and contrast. In the second panel the two 5 μm × 1 μm slots have been ion milled to a depth of 2.5 μm. The lowest panel shows output from the digital image correlation software, with displacements shown in units of pixels. In this case the displacements show that the stress is compressive.

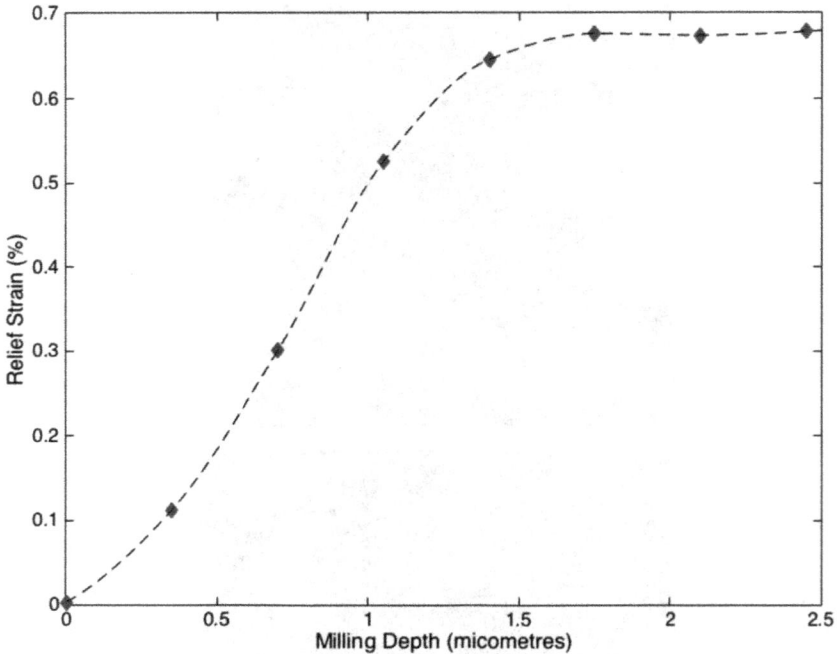

Figure 4.13. Plot of peak relief strain versus milling depth for the example shown in figure 4.12.

with respect to each other, other geometries of holes have also been used, including circular holes and ring-cores where an annular pattern is milled, leaving behind a circular post of material [26]. With these geometries it is possible to determine multi axial strain in a single milling operation. Beyond simple stress determination, it has also been shown that using this method it is possible to determine the Poisson ratio of a material via a two-stage milling process [30].

In addition to measuring stress via the strain relief that occurs on removing small volumes of material, FIB can also be used as an ideal tool to create marker systems to study larger-scale plastic deformation [31]. Ion-milled simple grids of either lines or dots on a regular pitch can be applied to the polished faces of specimens before they are plastically deformed. By imaging the grids after deformation it becomes possible study the deformation on the scale of the grain structure and see how grains slide past each during deformation. Such a technique can also be combined with EBSD to obtain the behaviour of individual grains of known orientations and how they deform with respect to their neighbouring grains [20].

4.6 Secondary ion mass spectrometry and atom probe

The surface science community has for many years been operating ion columns in parallel to the conventional FIB community with surprisingly little overlap of activity. Secondary ion mass spectrometry (SIMS) is a technique where a beam of focused ions is used to sputter a sample surface and the sputtered atoms are collected in a spectrometer synchronously with the sputtering process, so that the sample chemistry can be mapped and reconstructed [32]. Traditionally, SIMS instruments have been dedicated instruments and in most cases they operate with different ion species from FIB systems and also ionised clusters of atoms. Two different approaches to marrying the two techniques have been pursued. The first is the attachment of mass spectrometers to FIB instruments in much the same way an EDS detector might be fitted to an SEM, for example see [33] (chapter 13, in particular, and references therein). Instruments of this type are not currently commonplace as they are not as sensitive as dedicated SIMS instruments and so they are still somewhat limited in their use. The second type of instrument occurs with the addition of scan coils and imaging systems, such as secondary electron detectors, to SIMS column-based instruments, although the layout of the instruments means they are not optimised for conventional FIB operations and are still largely dedicated to SIMS.

FIB instruments are also used to produce samples for the atom probe microscope. With a common ancestry to the FIB, the atom probe was demonstrated half a century ago [34] and was derived from early field emission ion sources. Samples are essentially very fine needles of material and are evaporated by field emission with ejected ions being collected and identified to produce a three-dimensional reconstruction of the sample, ranging from a few thousand up to around ten million atoms, depending upon the specific behaviour of the sample and its preparation. In much the same manner as a TEM lamella is produced, the FIB will be used to cut and lift out a very thin needle of material with a tip radius of a few nm. Often the end of the tip will be made from one of the GIS deposition gases that was deposited as a small pad on the surface to reduce implant of the FIB Ga ions and the early part of the evaporation is discounted from the data; however, some Ga is still expected to be found in the sample. Owing to the small sample size, somewhat limited range of suitable samples and general complexity of the technique, the atom probe is still a relatively uncommon technique, but FIB does offer a route for producing samples.

References

[1] Williams D B and Carter C B 2009 *Transmission Electron Microscopy: A Textbook for Materials Science* (Berlin: Springer)

[2] Giannuzzi L A and Stevie F A 1999 A review of focused ion beam milling techniques for tem specimen preparation *Micron* **30** 197–204

[3] Munroe M P A 2009 The application of focused ion microscopy in the material sciences *Mater. Charact.* **60** 2–13

[4] Uchic M D, Groeber M A, Dimiduk D and Simmons J P 2006 3D microstrcutral characterization of nickel superalloys via serial-sectioning using a dual-beam fib *Scripta Mater.* **55** 23–28

[5] Borgh I *et al* 2013 On the three-dimensional structure of WC grains in cemented carbides *Acter Mater.* **61** 4726–33

[6] Groeber M A, Rowenhorst D J and Uchic M D 2009 Collection, processing and analysis of three-dimensional EBSD data sets *Electron Backscatter Diffraction in Materials Science* 2nd edn (Berlin: Springer)

[7] MacSleyne J, Uchic M D, Simmons J P and de Graef M 2009 Three-dimensional analysis of secondary gamma prime precipitates in rene-88 dt and umf-20 superalloys *Acter Mater.* **57** 6251–67

[8] Ni N, Lozano-Perez S, Sykes J M, Smith G D W and Grovenor C R M 2011 Focussed ion beam for the 3D characterization of cracking in oxide scales formed on commercial zirlo alloys *Corrosion Sci.* **53** 4073–83

[9] Korte S M, Ritter M, Jiao C, Midgley P A and Clegg W J 2011 Three-dimensional electron backscattered diffraction analyisis of deformation in MgO micropillars *Acter Mater.* **59** 7241–54

[10] Earl J S, Leary R K, Perrin J S, Brydson R, Harrington J P, Markowitz K and Milne S J 2010 Characterization of dendrite structure in three dimensions using FIB-SEM *J. Microscopy* **240** 1–5

[11] Mingard K P, Jones H G and Gee M G 2014 Metrolgical challenges for reconstruction of 3D microsctructures by focused ion beam tomography methods *J. Microscopy* **253** 93–108

[12] Holzer L, Indutnyi F, Gasser P, Munch B and Wegman M 2004 Three-dimensional analysis of porous BaTiO3 ceramics using FIB nanotomography *J. Microscopy* **216** 84–95

[13] Schaffer M, Wagner J, Schaffer B, Schmied M and Mulders H 2007 Automated three-dimensional x-ray analysis using a dual beam FIB *Ultramicroscopy* **107** 587–97

[14] Iwai H *et al* 2010 Quantification of SOFC anode microstructure based on dual beam FIB-SEM technique *J. Power Sources* **195** 955–61

[15] Principe E 2010 High density FIB-SEM tomography via real-time imaging US Patent Specification 0288925

[16] Jones H G, Mingard K P and Cox D C 2014 Investigation of slice thickness and shape milled by a focused ion beam for three-dimensional reconstruction of microstructures *Ultramicroscopy* **139** 20–28

[17] Uchic M D, Dimiduk D M, Florando J N and Nix W D 2004 Sample dimensions influence strength and crystal plasticity *Science* **305** 986–9

[18] Wang J, Yang G and Hodgson P D 2015 Mechanical behavior of nano-crystalline metallic thin films and multilayers under microcompression *Met. Mat. Trans.* A **46** 1404–12

[19] Kiener D, Grosinger W, Dehm G and Pippan R 2008 A further step towards an understanding of size-dependent crystal plasticity: *in situ* tension experiments on miniaturized single-crystal copper samples *Acter Mater.* **56** 580–92

[20] Shade P A, Groeber M A, Schuren J C and Uchic M D 2013 Experimental measurement of surface strains and local lattice rotations combined with 3D microstructure reconstruction from deformed polycrystalline ensembles at the micro-scale *Integrating Materials and Manufacturing Innovation* **2** 9772

[21] Barber A H, Lu D and Pugno N M 2015 Extreme strength observed in limpet teeth *J. R. Soc. Interface* **12** 20141326

[22] Kienera D, Zhangb Z, Sturmc S, Cazottesb S, Imrichb P J, Kirchlechnera C and Dehm G 2012 Advanced nanomechanics in the TEM: effects of thermal annealing on FIB prepared Cu samples *Phil. Mag.* **92** 3269–89

[23] Kienera D, Motzb C, Resterb M, Jenkoc M and Dehm G 2007 FIB damage of Cu and possible consequences for miniaturized mechanical tests *Mat. Sci. Eng.* **459** 262–72

[24] Baika S I, Choib E G, Junb J H and Kim Y W 2008 Defect structure induced by ion injection in WC-Co *Scripta Mater.* **58** 614–7

[25] Withers P J and Bhadeshia H K D H 2001 Residual stress: part 1. Measurement techniques *Mater. Sci. Technol.* **17** 355–65

[26] Korsunsky A M, Sebastiani M and Bemporad E 2009 Focused ion beam ring drilling for residual stress evaluation *Materials Lett.* **63** 2393–403

[27] Song X, Yeap K B, Zhu J, Belnoue J, Sebastiani M, Bemporad E, Zeng K and Korsunsky A M 2012 Residual stress measurement in thin films at sub-micron scale using focused ion beam milling and imaging *Thin Solid Films* **520** 2073–6

[28] Bemporad E, Brisotto M, Depero L E, Gelfi M, Korsunsky A M, Lunt A J G and Sebastiani M 2014 A critical comparison between XRD and FIB residual stress measurement techniques in thin films *Thin Solid Films* **572** 224–31

[29] Winiarski B, Gholinia A, Tian J, Yokoyama Y, Liaw P K and Withers P J 2012 Submicron-scale depth profiling of residual stress in amorphous materials by incremental focused ion beam slotting *Acter Mater.* **60** 572

[30] Sebastiani M, Eberl C, Bemporad E, Korsunsky A M, Nix W D and Carassiti F 2014 Focused ion beam four-slot milling for Poisson's ratio and residual stress evaluation at the micron scale *Surf. Coat. Technol.* **251** 151–61

[31] Rust M A and Todd R I 2011 Surface studies of region ii superplasticity of AA5083 in shear: confirmation of diffusion creep, grain neighbour switching and absence of dislocation activity *Acter Mater.* **59** 5159–70

[32] van der Heide P 2014 *Secondary Ion Mass Spectrometry: An Introduction to Principles and Practices* (Berlin: Springer)

[33] Giannuzzi L A and Stevie F A (ed) 2005 *Introduction to Focussed Ion Beams* (Berlin: Springer)

[34] Muller E W, Panitz J A and McLane S B 1968 The atom-probe field ion microscope *Rev. Sci. Instrum.* **39** 83–86

Chapter 5

Focused ion beam fabrication for metrology

5.1 Overview

As we have seen, the removal of material with FIB to study specimens and determine a host of properties is probably the main use for these instruments. However, the removal of material on the scales offered by FIB gives us the possibility to modify existing structures and devices to increase their sensitivity or resolution. This chapter describes some these devices, how they are modified and what they offer to metrologists.

5.2 Superconducting devices

SQUIDs are macroscopic objects operating in the quantum regime and are capable of measuring a wide range of physical parameters with unequalled sensitivity [1–4]; very few devices offer such a diverse range of uses. The natural and most straightforward quantity that a SQUID responds to directly is magnetic flux, although other quantities as diverse as spatial displacement, photon detection, or even mass detection are possible [3, 5–7]. The first SQUID devices developed were of relatively large macroscopic size (typically tens of micrometres or often more in linear dimension) [8]. It is really only recently that it has been appreciated that SQUID size may be radically reduced towards the nanoscale and that such devices will not only retain exceptional sensitivity but will, through their size, find applications in a whole new range of detection and measurement areas [9–13]. The parameters directly affecting performance of a SQUID are given in equation (5.1), where the minimum detectable energy change a SQUID can measure is described as

$$\varepsilon_n = \frac{\langle S_\varphi^2 \rangle}{2L} = 16 k_{\mathrm{B}} T (LC)^{1/2}. \tag{5.1}$$

Here ε_n is the minimum detectable energy change, where S_φ is the spectral noise density and L and C are the inductance and capacitance of the SQUID loop, respectively, the temperature is given by T and k_{B} is the Boltzmann constant. As this equation shows, there are two clear ways to improve the energy sensitivity of a SQUID loop. Clearly an improvement can be won by a reduction in the operating temperature. Alternatively, we can effect a similar improvement by reducing the inductance and capacitance of the SQUID loop through reducing its size. The FIB allows us to easily reduce loop dimension down to a submicron diameter, giving orders of magnitude of reduction in SQUID loop size in a very simple manner [14]. Reassuringly, it has also been clear for some time that the noise performance of SQUIDs may also be improved not only by reducing the operating temperature, but also by reducing the SQUID inductance and junction capacitance.

Restrictions on operating ever-smaller SQUIDs have arisen for several reasons. Firstly,'traditional' SQUIDs, incorporating Josephson tunnel junctions, are generally found to have dimensions greater than 1 μm, due to junction current-density limitations. They also require a trilayer deposition route, which needs an oxidation treatment. Secondly, tunnel junctions also possess significant capacitance, arising from their geometry. It has been recognised that the alternative use of microbridge junctions would very effectively reduce the junction size and capacitance, thereby allowing smaller SQUID inductance, exceptional measurement sensitivity and low noise performance. The devices fabricated by FIB can encompass both nanobridge junctions on large-loop SQUIDs and nanoSQUID loops and both have shown excellent performance [15], but it should be stated that some basic properties of the Josephson junctions are not fully characterised or completely understood. An example of a 50 μm SQUID with FIB-milled nanobridge junctions and a nanoSQUID is shown in figure 5.1.

Figure 5.2 shows the process of producing superconducting junctions schematically. The substrate is usually p-type 001 Si wafer with 100–200 nm of SiO_2 grown on the top side. An Nb film, usually 200 nm thick, is deposited on this by electron beam evaporation in ultra-high-vacuum. The film is usually patterned to produce device leads and bond-pads and region set-asides for the FIB to produce the

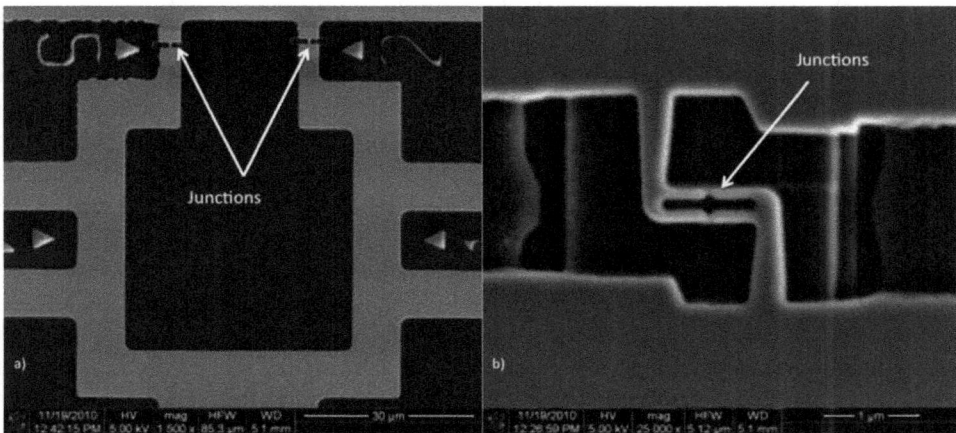

Figure 5.1. SEM micrographs of a 50 μm SQUID with nanobridge junctions (a) and a 1 μm × 100 nm nanoSQUID (b). Reproduced with permission from [14].

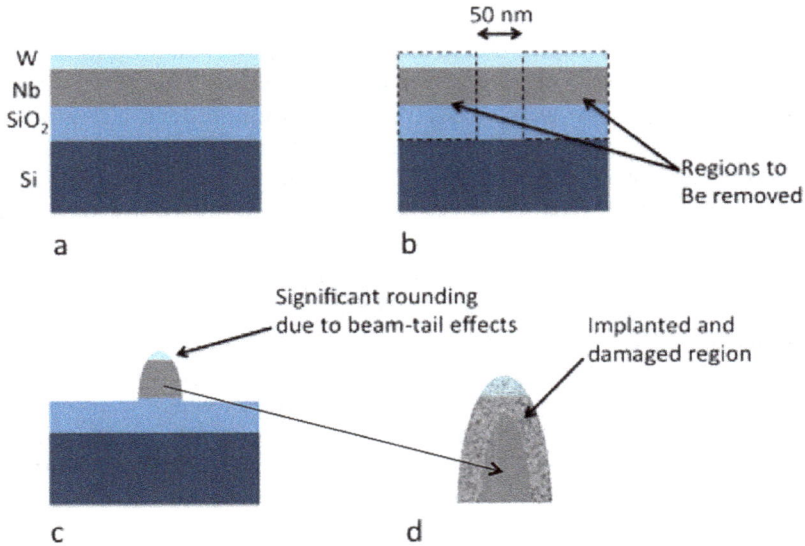

Figure 5.2. Schematic diagrams of the milling of weak-link superconducting junctions in Nb (not to scale). (*a*) Cross-section layout of the film. (*b*) Regions to be removed by FIB. (*c*) Actual finished junction geometry. (*d*) Close-up view of implanted and damaged region of finished junction. From [14].

SQUIDs themselves. On top of the Nb, where the SQUID will be made, is a thin film (50 nm) of electron beam deposited tungsten hexacarbonyl, put down by the electron beam in the FIB to protect the film from ion beam exposure when aligning to the region to be milled. Usually this is only deposited in the region where a SQUID loop or junction will be and is no more than a pad of 10×10 µm. To create the junctions a milling pattern is created that removes material from either side of the junction region. The final width superconducting junction is usually around 50 nm. The Gaussian profile of the beam spot is exploited in that when milling two adjacent regions with very small separation the beam tails overlap, leading to some milling in the area between these regions, which in turn leads to a rounding of the top of the remaining material. The ion milling creates an implanted and damaged zone in the outer regions of the milled junction. When the film and junction regions are electrically active there are several distinct features observed in their behaviour. Firstly, there are two superconducting transitions, the first, occurring at a higher temperature, is consistent with the non-milled film, but a second reduced T_c is also present and is consistent with the presence of a weak-link junction of reduced cross-section. It should be noted that these are not classical Josephson junctions of superconductor/insulator/superconductor or superconductor/normal-metal/superconductor (see [1] for descriptions of classical junctions).

FIB-fabricated Nb-based SQUID devices have a tendency to operate only one or two K below the superconducting transition of the film. It has been found that once a device is fabricated and the superconducting transition (T_c) of the junctions is established, the FIB can further dose the junctions to reduce the T_c. to the desired range. This process of course removes a small amount of Nb, but also moves the implanted and damaged front further into the junction. What this does in effect is reduce the size of the superconducting Nb grains in the junction region. It has been shown in both [16] and [17] that the superconducting transition in nanostructured Nb films can be lowered by reducing the grain size of the material. A monotonic reduction in T_c is observed with reducing grain size once it is smaller than approximately 28 nm, with the superconducting transition completely eliminated once the grain size is smaller than 8 nm. Based on the measured size of these junctions and knowledge of the implant depth for a 30 keV Ga ion, it is estimated the junctions are exactly of the dimensions to place them in the region where T_c is reduced. This simple mechanism is the underlying reason that enables the weak link junctions to function and within reason the T_c can be lowered by repeated doses in increments as low as 0.1 K, but it must be borne in mind that the inherent small variation in the number of grains in the junction region makes this at present a slightly imprecise method. The complete loss of superconductivity in high Ga ion dose Nb films can also be advantageous and is in fact one of the key aspects of how and why these devices are viable. The other very significant electrical feature seen with junctions of this type is the presence of a parallel shunt-resistor that is of vital importance to the operation of the devices. This shunt is the implanted and amorphised outer layer of the milled junction. When the superconductivity of the junction is broken by the high temperature, high current or the presence of high external field, junctions that are on the nanoscale have great difficulty in carrying the electrical current and surviving the subsequent Joule heating. The shunt in these FIB-fabricated junctions can assist in carrying this current loading when the junction loses its superconductivity, but the shunt is not superconducting and plays no role in normal operation.

The one aspect of FIB-fabricated nanoSQUIDs that is most outstanding is the noise performance of the devices. The measured magnetic flux noise spectral density S_Φ expressed in units of the flux quantum $\Phi_0/\mathrm{Hz}^{1/2}$, as a function of frequency from 0.1 Hz to 100 kHz, is shown in figure 5.3. Note that there is a region at low frequency where the noise spectrum has a $1/f^2$ form, but above 1 Hz, there is a much

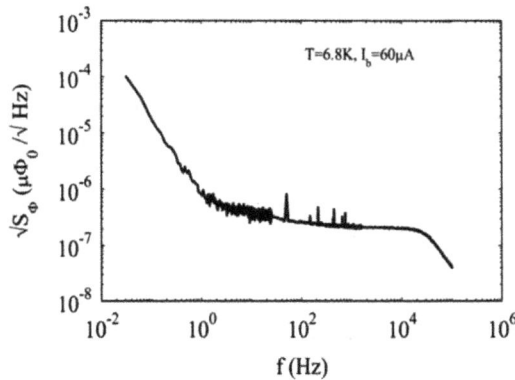

Figure 5.3. Flux noise spectral density versus frequency for a Nb nanoSQUID loop (350 nm diameter) at $T = 6.8$ K in zero magnetic field. Reproduced with permission from [15]. Copyright 2008, AIP Publishing LLC.

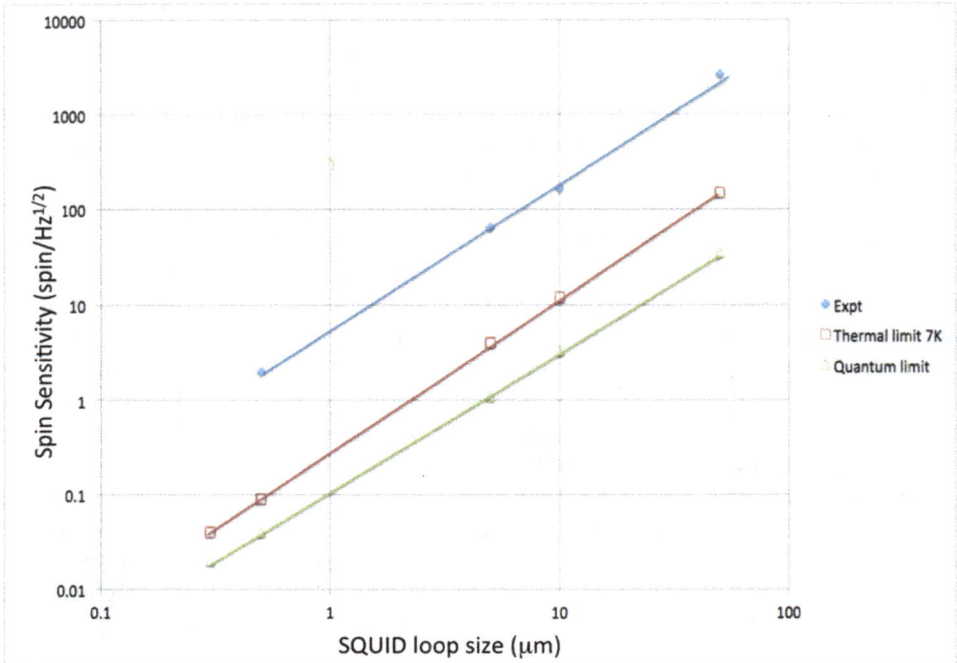

Figure 5.4. Spin sensitivity of a 300 nm diameter loop FIB-fabricated SQUID device. Reproduced with permission from [14].

weaker frequency dependence. Even at 1 Hz, the spectral density is as low as $0.8\,\mu\phi_0/\text{Hz}^{1/2}$, while in the white noise region around 1 kHz this has fallen to $0.2\,\mu\phi_0/\text{Hz}^{1/2}$. The frequency roll off at higher frequencies represents the result of filtering in the readout electronics. A more complete description of measurement and noise performance of FIB-milled Nb nanoSQUIDs is given in [15].

Given the extremely high sensitivity of FIB-fabricated nanoSQUIDs and super-conducting junctions, combined with the large range of measurements they are suitable for, they find many uses in the field of nano and quantum metrology. Analysing the sensitivity of FIB-fabricated nanoSQUIDs with a range of different magnetic particles (see section 5.3 for details on nanomanipulation) shows that the spin sensitivity of the devices scales with the loop size. Extrapolation of the perform-ance suggests that the ultimate limit of a 350 nm nanoSQUIDs might be as low as two spins/$\text{Hz}^{1/2}$ (figure 5.4). The fact that the sensitivity is still a significant margin above both the thermal limit at the operating temperature and the quantum limit suggests that with suitable care it may be possible to detect single spins with these devices.

5.3 Utilising manipulation systems

Owing to the widespread need to lift out TEM lamella that have been cut in FIB systems a large number of such instruments will be equipped with both a form of *in situ* manipulation system and a GIS to attach sharp tips to the lamella to transfer

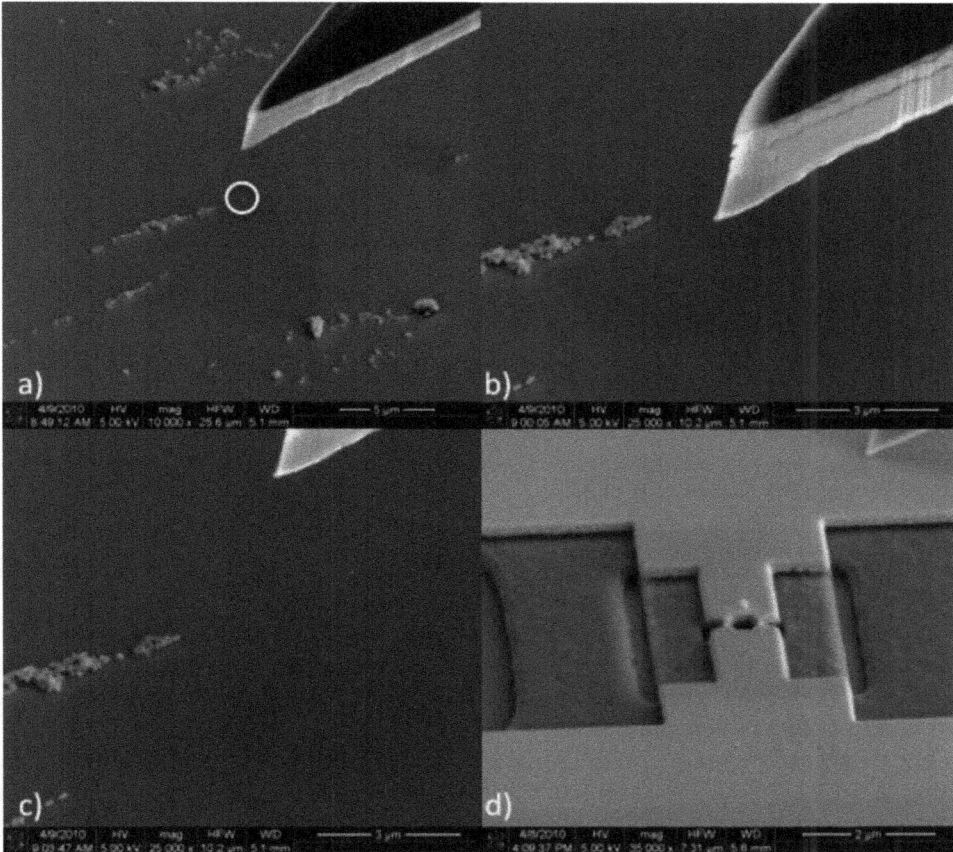

Figure 5.5. A manipulated Fe nanoparticle being placed on the edge of a nanoSQUID loop. (*a*) The dispersed particle on a separate substrate. (*b*) The sharpened carbon fibre tip attaching to a particle. (*c*) The particle is lifted from the substrate. (*d*) The particle is located on the edge of the SQUID loop. Reproduced with permission from [14].

them to copper TEM grids. Typically, manipulation systems will be based on piezoelectric positioners, with such systems offering tens of nm resolution (often better) and ranges of travel of several mm with at least three degrees of freedom. TEM samples can be temporarily attached to sharpened tips using the GIS gases moved to a TEM grid and removed from the tip after attachment to the grid by either cutting with FIB or pulling the tip from the lamella, now firmly attached in its new location. This form of manipulation can, of course, be applied to much more than TEM lamella and the methodology can be applied to construct devices from components or to move particles to devices to be measured. An example of this is shown in figure 5.5, based on the SQUID devices seen in the previous section, where the SQUID and the object to be measured must be in close proximity, often on a scale smaller than the size of the loop. In this example a separate substrate contains the magnetic nanoparticles and they are lifted from the sample by a sharpened

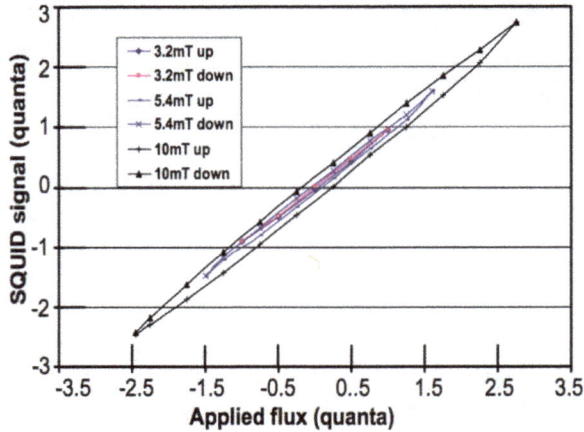

Figure 5.6. Hysteresis plots showing detection of the single FePt nanobead for a range of applied magnetic fields. Reproduced with permission from [14].

carbon fibre attached to a manipulation system. In this case electrostatic attraction is all that is required to lift the particles from the surface, where they are transferred to the SQUID loop. A short burst of Pt gas from the instrument GIS system, along with focusing the electron beam on the particle is all that is required to attach the particle to the SQUID loop. Figure 5.6 shows the effect of the bead on the nanoSQUID once exposed to applied magnetic fields.

The combined imaging, ion milling and GIS deposition and manipulation system found in FIB systems naturally lends itself to characterisation of specimens in the sub-micron size range along with the use of such small samples to test the resolution limits of other metrological techniques. Examples of devices where manipulated particles and structures are used to test these phenomena are given in [18–20] for micro Hall sensors, graphene Hall sensors [21] and nanomechanical resonance (via nanoSQUID) [22]. Other fundamental studies involving nanomanipulated magnetic particles can also be found in [23, 24]. Further examples of nanomanipulated objects are shown in figure 5.7

5.4 AFM cantilevers and dimensional artefacts for scanning probe techniques

A widely available instrumental technique used in many metrological areas is the AFM and its many derivative techniques. The resolution of most of the scanning probe techniques is dependent upon both tip geometry and instrument variables, such as the level of control of the piezo scanners used to move the tip. In most cases there is little scope to improve already existing instruments and it can be thought of as a limit that can only be bettered by improving the technology. However, an extremely simple way to improve the resolution of many forms of AFM is to improve the sharpness of the tip [25]. Although standard commercial AFM tips are manufactured to be sharp, and can achieve tip radii of a few nm, the sharp region is only at the very end of a

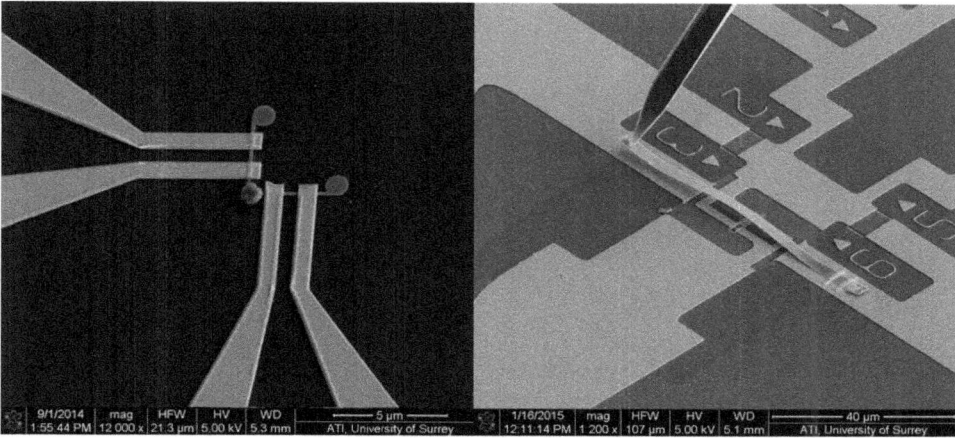

Figure 5.7. A magnetic particle that has been manipulated onto a magnetic domain wall device (left), and a micromechanical resonator being assembled above a SQUID device (right).

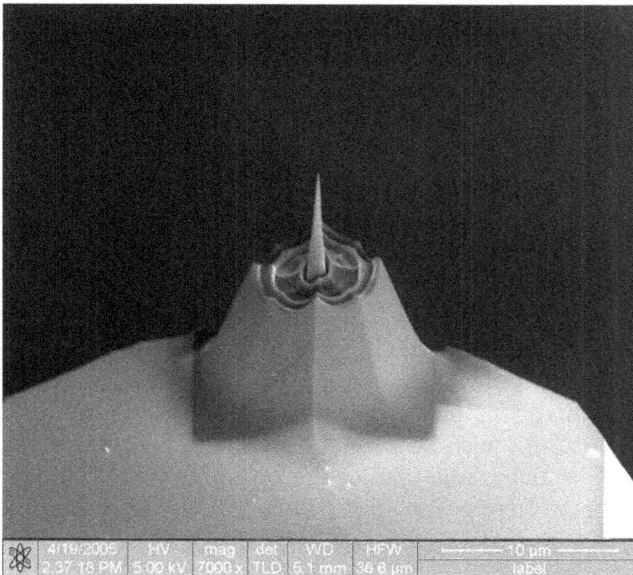

Figure 5.8. An FIB-sharpened AFM tip. The use of concentric annular milling patterns improves both sharpness and aspect ratio.

pyramidal tip. By ion milling a simple annular pattern it is possible to improve the sharpness and, equally importantly, the aspect ratio of the tip (figure 5.8).

Improving the sharpness of the tip improves the spatial resolution, allowing closer spaced items to be resolved, but improving the aspect ratio can also have a significant effect on resolution if the sample contains significant step edges, as when using conventional low aspect ratio tips the edge of the pyramid can detect the step edge

Figure 5.9. Comparison of imaging with standard and FIB-sharpened AFM tips. Left image is a topographic image of a calibration standard, right image is from the same standard (slightly different region) with a FIB-sharpened tip. A noticeable improvement in resolution is immediately apparent from the reduction in width of lines and lettering. Reproduced with permission from Andrew Yacoot (NPL).

before the tip itself, resulting in structures appearing wider than they really are. An example of this is shown in figure 5.9, where a comparison is shown between a conventional tip and a sharpened tip on imaging a standard calibration sample. The advantage of FIB-sharpened AFM tips has been demonstrated to the extent that several commercial AFM tip manufacturers offer them in their product catalogues, although unsurprisingly they do attract a price premium.

An additional example of where FIB has been used, along with other methods, to make unique scanning probe tips is given in [26]. Here conductive tips have had a Pt nanowire attached to the tip and are subsequently coated with an electrically and chemically insulating layer (parylene). The FIB modification of these tips is the most simple part of the process as the ion beam is used purely to remove the very end of the parylene coating, exposing the nanowire core. These hybrid tips are then used to probe local electrochemistry in fluids where reactions can take place only in the vicinity of the interaction volume of the tip. Another variation of a scanning electrochemical tip involving more extensive use of FIB is given in [27]. Significant modification of AFM tips has also been shown in [28], where the FIB has been used to make indenters, punches and stamps for nano imprint lithography, along with combined stamp and imaging tips on the same AFM cantilever.

Briefly revisiting the previous section (5.3), a further novel use of FIB for AFM tip modification is shown in figure 5.10. Here a conventional AFM tip and a portion of its cantilever is cut from the cantilever body and, using a manipulation system, it is moved to a quartz tuning fork. These tuning forks offer an alternative to conventional cantilevers [29] and find particular use in cryogenic AFM applications, where the conventional laser detection of tip motion can lead to unwanted heating of the system. The weakness of tuning fork AFM systems however is that the tuning forks, being derived from quartz timing crystals used on circuit boards, were never designed for AFM applications and have no tip. Tips can be added by simply gluing on a small piece of sharpened metal wire using adhesive, but this will alter the resonant frequency of the tuning fork in unknown and inconsistent ways and also have a negative affect

Figure 5.10. Manipulation of an AFM tip and portion of the cantilever (left) and its placement on a tuning fork (right) in a FIB system.

on the quality factor, making it more difficult to detect changes in frequency as the tip interacts with the sample. By using FIB and existing AFM tips it is possible to add a well defined and characterised tip to the tuning fork with so little mass that it has a barely detectable change in either frequency and quality factor [30].

The ability to remove volumes of material precisely also allows FIB systems to be used to manufacture artefacts that can be used for the calibration and testing of scanning probe instruments. Figure 5.11 shows a simple example of a test sample used for magnetic force microscope (MFM) calibration. Here a thin magnetic film of an Fe/Pt alloy is deposited on Si. The FIB has been used to create the observable pattern, where the film has been removed. Comprising a randomised pattern of binary squares of either film or no-film regions (in this case based on 500 nm squares), the MFM can be tested in both topographic and magnetic modes for precision, squareness of scan and resolution. The patterns can be reduced or increased in size or other geometric shapes can be included. However, in the case of magnetic samples, once the pattern contains very small features the film can lose its magnetism as the beam tails from the Gaussian beam begin to overlap, destroying the properties of the film through the ion damage cascade. Samples such as this can be used for testing many types of scanning probe technique, including, but not limited to topographic imaging, conductive imaging, MFM, electrochemistry and others, depending on the sample, film type and tip application.

5.5 Other devices

Such is the versatility of FIB in modifying structures by either sputtering or deposition processes that it lends itself to many areas of scientific study. Not only is it capable of purely metrological devices and structures, but it can be applied to many devices to study physical phenomena or test the feasibility of devices and techniques. At the time of writing, the literature contains in excess 1000 examples of partially or wholly FIB-fabricated devices and structures, with one particularly

Figure 5.11. MFM resolution and calibration sample. The pattern is milled into a Fe/Pt alloy film on a silicon wafer and is comprised of features based on random patterning of 500 nm squares occupying 50% of the pattern area.

strong area being the field of photonic and plasmonic devices. The main reason for this is not only the flexibility of the FIB instrument, but also the fact that the wavelengths of light typically used in these devices, and hence the sizes of structures they contain, are very much in the size range that is easily fabricable by FIB. For example, [31] demonstrated a FIB-fabricated two-dimensional photonic crystal, [32] fabricated plasmonic sensor components, and [33] employed both electron beam deposition using GIS and FIB milling to produce plasmonic waveguides.

Similarly, FIB has been used to produce a very novel plasmonic device by incorporating a waveguide fabricated on an AFM cantilever [34], and another FIB-modified AFM cantilever has been made into a fluidic dispenser capable of precise deposition of droplets as small as 200 attolitres [35]. A common feature of these two devices and many of the photonic structures is the use of FIB to introduce simple, but extremely precise, holes. Other more complex hole arrays can be introduced into suspended thin films that can in turn be used as lithographic stencils for patterning of nanoscale structures [36], where sub 50 nm wide metallic lines have been demonstrated.

References

[1] Clarke J and Braginski A (ed) 2004 *The SQUID Handbook Fundamentals and Technology of SQUIDs and SQUID Systems* vol 1 (New York: Wiley)
[2] Koelle D, Kleiner R, Ludwig F, Danster E and Clarke J 1999 High-transition-temperature superconducting quantum interference devices *Rev. Mod. Phys.* **71** 631–51

[3] Gallop J C 2003 Squids: some limits to measurement *Supercond. Sci. Technol.* **16** 1575–82

[4] Hilgenkamp H and Mannhart J 2002 Grain boundaries in high T_c supercondcutors *Rev. Mod. Phys.* **74** 485–549

[5] Veauvy C, Hasselbach K and Mailly D 2002 Scanning superconducting quantum interference device force microscope *Rev. Sci. Instrum.* **73** 3825–30

[6] Hilgenkamp H, Ariando Smilde H J, Blank D, Rijnders G, Rogalla H, Kirtley J and Tsuei C C 2003 Ordering and manipulation of the magnetic moments in large-scale superconducting pi-loop arrays *Nature* **422** 50–53

[7] Hao L, Macfarlane J C, Lam S K H, Foley C P, Josephs-Franks P and Gallop J C 2005 Inductive sensor based on nano-scale squids *IEEE Trans. Appl. Supercond.* **15** 514–7

[8] Awschalom D D, Rozen J R, Ketchen M B, Gallagher W J, Kleinsasser A W, Sandstrom R L and Bumble B 1986 Low-noise modular microsusceptometer using nearly quantum limited dc squids *Appl. Phys. Lett.* **53** 2108–10

[9] Lam S K H and Tilbrook D L 2003 Development of a niobium nanosuperconducting quantum interference device for the detection of small spin populations *Appl. Phys. Lett.* **82** 1078–80

[10] Cleuziou J-P, Wernsdorfer W, Bouchiat V, Ondarharcu T and Nonthioux M 2006 Carbon nanotube superconducting quantum interference device *Nature Nanotechnol.* **1** 53–59

[11] Gallop J, Josephs-Franks P W, Davis J, Hao L and Macfarlane J C 2002 Miniature dc squid devices for the detection of single atomic spin-flips *Physica* C **368** 109–14

[12] Hao L, Gallop J C, Cox D, Romans E, MacFarlane J C and Chen J 2009 Focused ion beam nanoSQUIDs as novel NEMS resonator readouts *IEEE Trans. Appl. Supercond.* **19** 693–6

[13] Troeman A G P, Derking H, Borger B, Pleikies J, Veldhuis D and Hilgenkamp H 2007 Nanosquids based on niobium constrictions *Nano Lett.* **7** 2152–6

[14] Cox D C, Hao L and Gallop J C 2014 Focused ion beam processing of superconducting junctions and SQUID based devices *Nanofabrication* **1** 53–64

[15] Hao L, Macfarlane J C, Gallop J C, Cox D, Beyer J, Drung D and Schurig T 2008 Measurement and noise performance of nano-superconducting-quantum-interference devices fabricated by focused ion beam *Appl. Phys. Lett.* **92** 192507–3

[16] Jin Y R, Song X and Zhang D L 2009 Grain-size dependence of superconductivity in dc sputtered Nb films Science in China Series G *Physics Mechanics and Astronomy* **1** 1289–92

[17] Bose S, Raychaudri P, Banerjee R, Vasa P and Ayyub P 2005 Mechanism of the size dependence of the superconducting transition of nanostructured Nb *Phys. Rev. Lett.* **95** 147003

[18] Kazakova O *et al* 2009 Detection of a micron-sized magnetic particle using InSb Hall sensor *IEEE Transaction on Magnetics* vol 10, pp 4499–4502 International Magnetics Conference 2009 (INTERMAG), 4–8, 2009 Sacramento, CA

[19] Kazakova O *et al* 2010 Ultrasmall particle detection using a submicron Hall sensor *J. Appl. Phys.* **107** 09E708

[20] di Michele L *et al* 2011 Detection and susceptability measurements of a single dynal bead *J. Appl. Phys.* **110** 063919

[21] Panchal V *et al* 2013 Epitaxial graphene sensors for detection of small magnetic moments *IEEE Transaction on Magnetics, IX European Magnetic Sensors and Actuators Conference* **49** 97–100 (EMSA)

[22] Hao L *et al* 2013 Coupled nanosquids and nano-electromechanical systems (NEMS) resonators *IEEE Trans. Appl. Supercond.* **23** 1800304

[23] Rod I *et al* 2009 The route to single magnetic particle detection: a carbon nanotube decorated with a finite number of nanocubes *Nanotechnology* **20** 335301

[24] Corte-Leon H, Krzysteczko P, Schumacher H W, Manzin A, Cox D, Antonov V and Kazakova O 2015 Magnetic bead detection using domain wall-based nanosensor *J. Appl. Phys.* **117** 17E313

[25] Yacoot A and Koenders L 2008 Aspects of scanning force microscope probes and their effects on dimensional measurement *J. Phys. D: Appl. Phys.* **41** 103001

[26] Wain A J, Cox D C, Zhou S and Turnbull A 2011 High aspect-ratio needle probes for combined scanning electrochemical microscopy atomic force microscopy *Electrochemistry Communications* **13** 78–81

[27] Lugstein A, Bertagnolli E, Kranz C, Kueng A and Mizaikoff B 2002 Integrating micro and nanoelectrodes into atomic force microscopy cantilevers using focused ion beam techniques *Appl. Phys. Lett.* **81** 349–51

[28] Menozzi C, Calabri L, Facci P, Pingue P, Dinelli F and Baschieri P 2007 Focused ion beam as tool for atomic force microscope (AFM) probes sculpturing *J. Phys.: Conf. Ser.* **126** 012070

[29] www.cdn.intechopen.com/pdfs-wm/36329.pdf

[30] Rajkumar R 2014 Novel nanoscale electronic devices for metrology *PhD thesis* (University of Surrey)

[31] Paraire N A, Filloux P G and Wang K 2004 Patterning and characterization of 2D photonic crystals fabricated by focused ion beam etching of multilayer membranes *Nanotechnology* **15** 341–5

[32] Madisona A, Dhawana A, Gerholda M, Vo-Dinha T, Russella P E and Leonarda D N 2009 FIB/SEM fabrication of nanostructures for plasmonic sensors and waveguides *Microscopy and Microanalysis* **15** 354–8

[33] Dhawan A, Gerhold M, Madison A, Fowlkes J, Russell P E, Vo-Dinh T and Leonard D N 2009 Fabrication of nanodot plasmonic waveguide structures using FIB milling and electron beam-induced deposition *Scanning* **31** 139–46

[34] Park I-Y, Kim S, Choi J, Lee D-H, Kim Y-J, Kling M F, Stockman M I and Kim S-W 2011 Plasmonic generation of ultrashort extreme-ultraviolet light pulses *Nat. Photonics* **5** 677–81

[35] Meister A, Krishnamoorthy S, Hinderling C, Pugin R and Heinzelmann H 2006 Local modification of micellar layers using nanoscale dispensing *Microelectron. Eng.* **83** 1509–12

[36] Santschi C *et al* 2009 Focused ion beam: a versatile technique for the fabrication of nano-devices *Praktische Metallographie-Practical Metallography* **46** 154–6

Chapter 6

Future developments

6.1 Where we currently are

As has been demonstrated, there are few scientific instruments currently in use today that contribute to such a range of applications as FIB. It is even rarer for an instrument to be able to both make measurements and create samples and devices that will be measured or make measurements on other instruments. A large selection of the *in situ* measurements do, of course, rely on other equipment attached to the system, with the now very commonly installed SEM column with its usual accompaniment of imaging detectors and additionally EDX and EBSD detectors chief amongst them. At the same time, chamber sizes have on average grown and instrument stages have followed suit, enabling the installation of *in situ* mechanical test rigs, indenters, probe stations, hot and cold stages and even atomic force microscopes. Indeed, FIB has developed from the prototype stage in just one or two laboratories into a widely available commercial instrument available from several manufacturers at the same time as techniques in the SEM have been developed and matured. Many of these techniques, such as EDX, EBSD and even imaging (via in-lens detectors) have developed to full maturity for two-dimensional analysis in the SEM. What FIB has done is open the door to enable measurement in the third dimension for many of these techniques.

However, the one great strength of FIB—the removal of sections of material—is also the feature that is most prone to error. To clarify this statement, as a general rule, providing the sample is well behaved, i.e. not charging and causing beam drift, or alloying as in the case of Sn with Ga ion beams, a patterned area would be expected to be accurate to within a few per cent. The desired location and size of the patterned area will be in excellent agreement with the chosen values. This, of course, will depend to some extent on beam currents and pattern size, but the experienced user would choose appropriately and this should hold true. This is ably demonstrated by examples such as those used in residual stress measurement or milling nanoSQUIDs, or where the objective is just to make something as sharp or thin as possible, for example AFM tips or TEM lamellae. The third dimension, the depth, is of course much more difficult to control, as the density and crystallinity will greatly affect the sputter yield, resulting in differing depths for a given ion dose.

The main sources of errors arise in the FIB milling process when multiple slices are milled sequentially for either imaging, EDX or EBSD, with these then being reconstructed as representative volumes in software. It has been demonstrated that these errors can be significant, and often indeterminable. There are many examples in the literature of three-dimensionally reconstructed volumes using some or all of these techniques and only on extremely rare occasions is the subject of possible errors in the reconstruction mentioned. What we can be sure of is that they almost all do contain errors, except in cases where the structure was known in advance, somewhat negating the usefulness of the experiment. Correcting or even establishing these errors is not a trivial task, but it is one that needs addressing if FIB is to be considered a serious tool for three-dimensional metrology.

Two strands need to be considered in establishing FIB as a reliable tool for three-dimensional reconstruction. One is that it is not clear what the source of these errors is. For example, is it due to stage movement or drift, or could it be that the algorithms used to track fiducial markers are not robust, or is it that instabilities in the beam are causing it not to scan where it was intended? To a large extent these are manufacturer issues and only they can solve them on commercial instruments. Where metrology laboratories can help out is in creating standards of known internal geometry so that errors can be established and improvements in both software and hardware can be implemented. Metrology laboratories can also add extra equipment not normally available to the regular FIB user, such as laser interferometers to stages and samples to check stage accuracy and provide feedback to the FIB community on sources of errors. The second issue can be addressed by many of the users themselves by adding marker systems to their sample surfaces, if at all possible, so that errors can be spotted and at least partially determined during the reconstruction. Not attempting to establish these errors casts significant doubt on the validity of the measurements that users produce and they should be treated with due caution where there is no mention of error and there has been no attempt to determine it.

6.2 The end of the Ga ion source?

The vast majority of FIB systems around the world have Ga ion sources, so why would we ask if Ga has a future? To partially answer this; quite simply, Ga will continue to be the dominant source for some considerable time due to the large number of instruments already in the field and with considerable service lives remaining. However, what is clear is that Ga is not the best source for imaging (in resolution and sample damage terms) and is also not the best source for high volume sputtering. At the imaging end of the spectrum the gas-field He source offers a resolution improvement of around a factor of ten, while at the same time imparting very little damage into the specimen. For high volume sputtering, the plasma Xe source offers massive gains in sputter rates, it must be said not due to the mass of Xe but because of the higher current available in the plasma source.

Replacing high resolution SEM with an He ion FIB is still a growing field. However, the relatively high cost and complexity of the gas-field source instruments dictates that at present this is not a solution for all users. Furthermore, no-one would have purchased an He ion FIB as an alternative to a Ga instrument, and in this context suggesting He as an alternative to Ga is a little unfair. However, the additional benefits of these sources when running higher mass species such as Ne allows modest sputter yields and brings the instrument to the attention of the nanofabrication community, clearly overlapping with, and to some extent, bettering the performance of Ga. For those very high performance imaging and sputtering operations it may be that Ne is a viable alternative, but for more general use it will still not have the widespread applicability of the higher mass and higher current sourced machines.

The relative newcomer to the commercial FIB community, the Xe plasma source machines, clearly have a much larger overlap with Ga instruments. At the upper end of the current range, plasma sources, while not having the brightnesses of a Ga LMIS, offer much more potential to increase sputtered volume and offer significant benefits in throughput and increase the range of samples and materials that can be studied. An obvious metrological area where this would have significant benefit is in the fabrication of test pieces for micromechanical testing. As has been shown with the maximum realistic sizes of test piece that can be fabricated in a Ga FIB, the samples are still of a size where very significant size effects are taking place and the sample is acting in ways that are not representative of the bulk. At the same time, the ion implant damage range is a significant portion of the sample size, causing other effects that are again not representative. The plasma source allows samples to be made at a size where size effects do not play a role, while at the same time the stopping range of Xe is also lower than Ga, reducing the depth of the damaged layer.

To summarise, there is still significant life ahead for the Ga LMIS at the time of writing. However, should a way be found to increase the brightness of the plasma source to enable it to operate with few nm beam spot sizes in the pA current range, without sacrificing the already astonishing performance in the μA range, then Ga will have very significant competition.

6.3 Final thoughts

From its first arrival as a commercially available machine, the FIB has always been a metrological instrument. The early adopters, chiefly the semiconductor community, used the instruments to check their production processes and to fault-find when things had gone wrong. Today, a very large number of laboratories have FIB systems involved in many areas of research, almost all of them using the instruments as the basis for making measurements. Some are used almost exclusively for preparing TEM lamellae, while others find very diverse applications and study many different materials. The one common theme that links them all is the ability of these instruments to make, cut and explore on the scale from a few nm to several hundred μm. This dimension range and versatility is covered by no other instrument currently available, and the constant improvement in performance from the manufacturers combined with the inventiveness and ingenuity of the user community will see them at the cutting edge of many metrological applications for the foreseeable future.

www.ingramcontent.com/pod-product-compliance
Lightning Source LLC
Chambersburg PA
CBHW081553220326
41598CB00036B/6663